STRANGFORD LOUGH

The Wildlife of an Irish Sea Lough

TO FRANCES

Remembering those happy times when we explored distant shores

Strangford Lough

The Wildlife of an Irish Sea Lough

by

ROBERT BROWN

The Institute of Irish Studies

The Queen's University of Belfast

1990

Published 1990

The Institute of Irish Studies,

The Queen's University of Belfast,

Belfast

HB ISBN 0 85389 355 1
PB ISBN 0 85389 356 X

© Robert Brown

All rights reserved. No part of this publication may be reproduced, stored in a retrieval system, or transmitted in any form or by any means, electronic, mechanical, photocopying, recording, or otherwise, without the prior permission of the copyright holder.

Printed by W. & G. Baird, Ltd., Antrim
Cover design by Rodney Miller Associates

Contents

Acknowledgements		vi
Chapter 1	Introduction	1
Chapter 2	How the Lough was Made	7
Chapter 3	Tides, Winds and Waves	33
Chapter 4	Life Below the Waves	53
Chapter 5	On the Shore	83
Chapter 6	Above the Shore	111
Chapter 7	The Wintering Birds of the Shore	139
Chapter 8	Nesting Birds of the Lough	167
Chapter 9	Postscript	195
Gazetteer		205
Some Books Recommended for Further Reading		219
Index		221

Acknowledgements

I have known Strangford Lough since 1970, when as a raw student I went on field courses at the Queen's University Marine Biology Station in Portaferry. There I was first introduced to the Lough...and to the delights of various hostelries in the village. Since then, apart from some time spent abroad, I have been lucky to have been able to spend a lot of time exploring the Lough and its incredible wealth of wildlife. My work and my leisure have taken me around the Lough, on it, under it, and over it, and I know that however long I continue to do so, there will always be new things to learn and discover, and old assumptions to be revised.

Looking back over all these years, and my time spent working with the National Trust, one of my strongest and happiest recollections will always be the time spent out on the Lough with many other people who know and love its wildlife - sadly some are no longer with us. We have all spent long evenings exchanging ideas and information, and discussing many aspects of the Lough or its wildlife. A number have spent time with me braving the bitter elements, when the Lough has been in one of its darker moods. There are many such people, and I have learned so much from them. Collectively, they and others in related fields of knowledge form a rich storehouse of knowledge and experience into which I have delved freely, and they have all in their various ways helped me write this book. Any success it may have will be largely due to them, though of course I hold responsiblity for any errors that have escaped attention.

My family have been a marvellous source of encouragement during the writing of this book, and I am deeply grateful to them, particularly for their wonderful encouragement through some very difficult times. I would also like to thank all those who have kindly allowed me to use their photographs; I believe they help to convey the excitement and beauty of the plants and animals of the Lough as well as any text.

When writing this section I wondered how to mention and thank all these good people, because there are a lot of them! In the end I have taken the easy alphabetical option. I have however, highlighted those patient souls who have read and checked various drafts of chapters '*', and those whose photographs I have used '‡', their initials are indicated in the picture captions as appropriate. To them all, I am particularly grateful for their generous contribution of knowledge, time, and encouragement:

Dave Allen*, Bernard Anderson, Dave Andrews*‡, Brian and Lesley Black, Bob Bleakley*‡, Pat Boaden*, Joe Breen, Jon Brook‡, Donald and Dinah Browne, Thanny Brownlee, Ronnie and Gwen Buchanan*, David Cabot, Hans Carse, Paddy and Anne Casement*, Ali Davidson, C. Douglas Deane‡, Neill Donaldson, Phil Doughty*, Norrie Duggan, David Erwin*, Andy Ferguson, Jeff Fleming, Ian Forsythe, Tony Fox, Joe Furphy, Vivian Gotto, Jack Gray, Julian Greenwood*, Tony Griffiths*, John Harwood*, Barrie Hartwell*, Peter Hawkey, Arthur Irvine, Palle Jepsen, Pat Jess, Alan Johnston‡, Philip Johnston, Tony Lord*, Paddy and Julie Mackie*, Jesper Madsen, George Magowan, William Marshall, Will McAvoy*‡, Peter McBride, Noel McCann, Christine McCawl‡, Carey and Sue McClay, Grahame McElwaine, Neville McKee, George McMurray, Oscar Merne, Meteorological Office (N.I.), Bob Milliken, David Moore, Mike Moser, Robert Nash, Micheal O'Briain, Ray O'Connor, Julian Orford, Gareth Owen, Reg Parker, Karl Partridge, Bernard Picton‡, Eric and Sheilagh Rainey*, Eddie Regan, Billy Reid, Dai Roberts, Desy Rojers, Sally Rousham*, Grahame Savage, Peter Scott, Ray Seed, Leslie Simpson*, Mick Taggart, Jack Torney, Noel Taggart, Lance J. Turtle‡, Brian Walker, Richard Warner*, Jo Whatmough*, Derek* and Vivian White, Richard Whyle, Christian Wiborg, Bertie Wilkinson, Peter Wright.

Finally, I and the publishers would like to express our gratitude for the generous support given to this book by the Department of the Environment (Countryside and Wildlife Branch), the Esme Mitchell Trust, and the National Trust.

1 *Introduction*

A LONG arm of the Irish Sea reaches deep into the heart of County Down. Strangford Lough, one of the largest sea loughs in Ireland, and certainly the most complex, is no ordinary inlet. In the Narrows flowing between the fishing villages of Portaferry and Strangford, or the waters lying between the rolling hills of Killyleagh, Whiterock and Kircubbin, and on the wide mudflats off the market towns of Newtownards and Comber, lies some of the richest marine and coastal wildlife, not just in Northern Ireland, but in the whole of Europe.

The Lough is some 150 square Km in area, and over 30Km in length. Its origins go back about five hundred million years, and the rocks and sediments deposited since then have had a major role in creating the character and topography of the Lough that we see today. In some places it is up to 60m deep, in others it is studded with islands and shallow reefs, or fringed by mudflats, marshes, rocks, bays and headlands. Its character is equally variable. At times it has a peaceful, almost glass-like quality, echoing its Celtic name, Cuan - 'the quiet lough'; on other occasions, its dark waves and powerful tidal currents fully justify its Viking name, Strangford - 'the strong fiord.' But within this landlocked area of sea, a unique combination of factors has resulted in riches that far exceed the dreams of the Vikings.

All coasts have their share of marine life, birds, coastal vegetation and scenery, and we can appreciate how the wildlife varies with the changing character of the coastline. In Strangford Lough these components are not only present in great variety, but also in great abundance. The Lough's waters, although part of the Irish Sea, harbour far richer life than would be found in an equivalent area outside the Lough. Above the waters, the shores of ancient rocks and glacier-deposited materials change their character with every new aspect. The islands, some gently rounded and fertile, others bleak and exposed, also reflect the riches of the Lough.

The Lough's wildlife is recognised internationally for its importance, and belatedly within Northern Ireland there is also a developing awareness of its value. No comparable area in Europe has such a range of habitats supporting such luxuriant life. The marine life is so dense that in many areas it is impossible to see the bottom or the shore for the number of plants and animals growing there, many of them extremely beautiful, others quite grotesque. There are several types of marine mammals, most notably one of the largest concentrations of Common Seals (*Phoca vitulina*) in Ireland, but also Grey Seals (*Halichoerus grypus*), Porpoises (*Phocoena phocoena*), and occasionally other types of whales.

Birds, many in flocks of thousands, others as solitary predators, come and go with the seasons, and in doing so they link the life of the Lough to many other areas of the world. Strangford's bird populations play an international role by virtue of their migrations. The Lough is one of the most important areas for birds overwintering in Ireland, and it holds over two thirds of the whole west European population of Pale-bellied Brent Geese (*Branta bernicla hrota*) on their arrival from their breeding grounds in Arctic Canada. Numbers of other wildfowl and waders are also extremely important, and these may spend their winters on the Lough or use it as a staging post in a complex pattern of movements about the northern hemisphere. After the wintering birds have left, other species move in from the south to nest on the many small islands. Some of the colonies are enormous, crowded and noisy, whilst other nesting birds are secretive and solitary.

The islands have their own wealth of wildlife. No two islands are the same: some are robust outcrops of bedrock, others mere slips of shingle and sand, and each has its own range of habitats and species. However, many share the common feature of having been protected from a number of human activities in the last century, and as a result they are

---

NAMES OF SPECIES

Most species possess two or more names; there may be several local, or colloquial names. 'Gorse' and 'Whin', for example, are the same species of plant. Unfortunately a few species, particularly marine ones, have not acquired such names. There is also a scientific name for every known species - in the case of Gorse/Whin, this is *Ulex europaeus*, whilst the rarer Western Gorse is *Ulex gallii*.

In this book, the first time a species is mentioned, both the colloquial and scientific names are given. Thereafter, only the colloquial name is used, unless the species does not have one, in which case the scientific name is retained. Those dipping into the middle of the book and wishing to find the scientific name of a species can do this by looking in the index; the first page reference to a species will generally be that containing both names.

richer in wildflowers with their attendant insects and other animals than most areas on the mainland.

All these different features of the Lough are not a series of disconnected areas each with its own share of inhabitants. They are linked by the movements and interactions of its wildlife, and by one fundamental feature that is both obvious and complex; the water. The powerful tidal currents of the Narrows, penetrating far in to the more sheltered reaches of the Lough, are the life blood of the whole system. Water movement, by currents and waves both maintains and slowly modifies the physical structures of the Lough by eroding, sorting, and depositing the materials. It also provides the rich supply of nutrients and microscopic life that feeds larger animals and plants; it transports the larvae that will eventually settle and grow into adult animals living on the bed of the Lough and its shores. Water movement therefore supports the marine life that is such a rich source of feeding for birds, and even some terrestrial animals, in both summer and winter; it creates new habitats for both nesting birds and coastal plants. Thus the whole Lough acts as a single entity, and a major theme of this book is how the different habitats and the different species interact with each other, often in ways that we find unpredictable.

For us humans, one of the most exciting features of the Lough, and one of its greatest challenges, is its accessibility. The Lough is not some remote wilderness in a distant corner of the world, which, apart from the lucky few, most people will only see on T.V. documentaries. It is in a densely populated country, close to a major industrial city, whose inhabitants increasingly have both the time and the resources to get out and enjoy their countryside. A network of country roads are there to carry even the least adventurous to most of the Lough shores, and having a boat to venture onto the Lough's waters is well within the budgets of many people. Pleasant villages and towns with all the usual amenities are scattered about the Lough. At the same time there are other parts of the Lough which are more inaccessible, and even dangerous in places, attracting those who seek physical challenges and solitude, and who want to encounter wildlife and habitats that show few signs of any human presence.

The challenges that come from the Lough's accessibility are many and varied, but they can be distilled into one key issue. The wildlife that we enjoy today, and which has the potential to play a great role in Strangford's attractions as an amenity area, is fragile. We do not understand completely how it works, and it is unlikely we ever will. But we do know enough to be able to act responsibly in the way we use the Lough. At present we are not using this knowlege to its full extent, and the future of the wildlife and the beauty of the Lough is at risk. If we want future generations to get the same excitement, interest, and leisure from the Lough that we now enjoy, it is time to plan now, to ensure that the wildlife and human society will still live together in the future.

This has many fundamental implications for the way we need to look after the Lough, but even on a small scale, each one of us has a small but important role to play in this respect. If this book has any success, it will result in more people visiting the Lough and enquiring into its riches of wildlife, and the more people get pleasure and excitement from doing this, the greater the impetus for the Lough's conservation. This does carry a risk however, since exploring the Lough inevitably carries the possibility of disturbing its wildlife; even simply turning over a boulder may disrupt hundreds or thousands of animals living beneath. By remembering certain precautions when we explore the Lough, such effects will be kept to a minimum, we will get more enjoyment from the Lough, and its wildlife will remain to play its role in the Lough, and will be there for others to enjoy:

* If you are diving, try to avoid collecting live animals. Select seashells which are obviously empty. If you plan to obtain a few shellfish to eat, only collect the largest (that should come naturally!) and leave some for the next person... it could be you!

* On the shores, do not collect very delicate and fragile animals; they will probably die unless you have specialized equipment. Preferably, return all specimens alive to the sea when you have examined them. Replace any rocks you have turned over, the right way up, and fill in any holes you have dug

* On the islands and mainland, it is illegal to remove plants without the owner's permission, and it is better not to pick the flowers - look, photograph or draw, then leave them for others to enjoy.

* Avoid disturbing the birds and other animals anywhere. Be still and quiet. Often they will then come near you, and as they are then behaving naturally, what you see will be much more interesting. Especially, leave nesting birds alone - they and their young are at their most vulnerable at this time. Nesting islands are marked with signs asking people not to land, but there are many other islands available for visiting.

* There are many areas with special designations and laws made to protect their wildlife. Many can be visited, and if you

Introduction 5

> need further information, contact the Department of the Environment (Countryside and Wildlife Branch), the National Trust, R.S.P.B., or the Ulster Wildlife Trust.
>
> * There are more details about organisations and sources of information elsewhere in the book.

Finally a few words about the book. It begins by looking at the ancient history of the Lough, its formation, and the role its underlying structures play today. Then the tides, currents, winds and waves are examined, particularly with respect to its structure and shape, and the different materials that make up the shores, sea bed, and islands. Having set the scene by considering the physical aspects of the Lough, different chapters explore the wildlife to be found in the waters and on the bed of the Lough, and on the shore. After this the life above the shores and on the islands and mainland is examined. Then there is a shift in emphasis; by contrast with the major environments to be found about the Lough, the next two chapters concentrate on the Lough's birds, firstly those arriving to spend their winters feeding on the shores, and secondly those that nest in the Lough during spring and early summer. This is followed by a brief chapter in which I make some tentative and extremely general conclusions about the way the Lough works as a system, how it is constantly

Shanny or Common Blenny (Blennius pholis) *revealed under a boulder near Portaferry. It is important to replace the boulders, or many species die from dessication.*

changing, and how we should ensure that its riches remain for future generations of plants and animals, including humans, to enjoy. The book ends with a gazetteer of some of the sites to be found about the Lough that have particular relevance to the natural history of the Lough.

2 *How the Lough was Made*

TAKE A walk along any shore of Strangford Lough, and look at the rocks, pebbles, and finer sediments. Different colours and textures greet the eye, blending with those of the plants growing nearby. Grey gritstones, shales and slates, almost black when wet, pale when dry, and patterned with white quartz veins, contrast with the volcanic materials, the speckled granites, and the deep red-browns of some sandstones, gold of others. The sediments vary in colour too; the sands in the north east of the Lough have a distinctly reddish tinge compared with those only a few miles further south. The rock formations show equal variety, with hard edged and folded beds contrasting with the rounded and crumbling outcrops of sandstone on other shores. Whole patterns of landscape reflect these different materials. Rugged scenery in the far south of the Lough contrasts with the almost feminine curves of the hillsides round Killinchy, whilst the bold outcrop of Scrabo Hill hangs over flat lowlying farmland round Newtownards.

It does not take long to realise that the landscape and its component materials have been subjected to a great variety of different processes. In fact, the present appearance of these rocks and sediments, many of which are extremely ancient, are due as much to what has happened to them since their creation, as it is to their original ingredients. By looking at the materials that make up the shores and the surrounding landscapes, and understanding what has happened to them, we can get a dramatic glimpse of the way in which the Lough itself was formed, moulded, scraped and deposited on, over millions of years. It is an epic story, still continuing, and one that still is not fully understood. It reaches back, not only into the ancient prehistory of Ireland, but of the world itself, when the continents and the mountains were being formed.

In The Beginning ...

Five hundred million years ago, a deep ocean covered a large section of the earth's crust, including the part that we now call County Down. To the north, where our Counties Antrim and Londonderry would be, lay the eastern edge of a large continent that extended from the places now occupied by Spitzbergen through to Greenland and Canada. The climate was tropical; not only was the world generally warmer, but the section of the earth's crust that concerns us straddled the equator at the time, and only later drifted north to its present position. Conditions may have been superficially similar to some of today's tropics. The marine life near the coast was a rich combination of corals (much as in modern tropical seas, but of a different type), sea lilies, relatives of the Feather Stars (*Antedon bifida*) in the Lough today, primitive armoured jawless fishes, and locally dense populations of brachiopods, or 'lamp shells.' Trilobites, like large woodlice and related to the horseshoe crabs still found in many parts of the world, were abundant.

Further away from the coast, as in modern seas, the marine life was less abundant, whilst the dark, tranquil depths would have allowed for the deposition of sediments. Doubtless there was a large supply of this material from rivers flowing off the continent. The coarser grits and sands settled close inshore, whilst the finer particles drifted some way out before finally settling on the bottom. This situation, even in modern seas, sometimes causes submarine 'landslides,' when coarser material accumulated near the coast slips down to cover the finer material that has previously accumulated in the depths. Thus layers of fine silt were suddenly overlain by much coarser material, and as this process was repeated, it created a series of horizontal layers, or beds, of varying consistency. These gradually consolidated into rock, forming beds of coarser greywackes (named from the German 'grey rocks' in the Harz Mountains), and fine shales or mudstones, with various intermediate types reflecting local conditions. With deposition continuing for perhaps a hundred million years, great volumes of these rocks accumulated and some areas of the sea began to shallow into a series of lagoons.

This long process continued through two major geological periods referred to as the Ordovician and Silurian. In keeping with their origin in the relatively lifeless depths of that ancient sea, the rocks hold few fossils although geologists have found occasional tracks thought to be left by bivalves and snails, and the fine skeletons of graptolites, small drifting relatives of our modern hydroids and sea anenomes. However, much was to happen to these ancient grits and mudstones before they aquired their present form.

A period of massive upheavals accompanied by violent volcanic eruptions brought this period of deposition to a close. Shifts in the earth's crust crumpled land masses and rock formations, forcing up mountain ridges that we still see in the southern uplands of Scotland, the English Lake District, and the mountains of north Wales. The former of these areas gave its name to this mountain building period,' which is referred

How the Lough was Made

Compressed and folded Silurian grits and shales at Killard Point. These rocks underlie most of the Lough and its glacial deposits.

to as the Caledonian. In what is now County Down, the grit and shale deposits were tightly compressed as large sections of the earth's crust were distorted under the strain. In places these pressures partially modified, or metamorphosed, the finer sedimentary materials of shale and mudstone to produce narrow beds of a rough slate. Elsewhere, massive fractures called faults, rent the rocks, and there are many such faults running through the Ards Peninsular, Lecale and the western shore, as well as under the Lough itself. These, together with innumerable smaller breaks, called joints, have been etched out by erosion, an excellent example being the deep cleft, called 'the Gurgle-gurgle' or Burney's Hole, at the tip of Killard Point. Everywhere, the Silurian rocks were folded and contorted.

During this period of deformation the rocks, which were probably thrown up into mountain ranges, in some places melted, producing upwellings of molten material and hot gases. Locally where the former cooled and hardened it formed the granite (more accurately, granodiorite) outcrops extending from near Newry to Slieve Croob near Ballynahinch. A number of narrow veins, or dykes, of pale greyish rock called lamphophyre were also formed, as molten rock was forced through small cracks and weaknesses in the bedrocks over a wide area.

These ancient, contorted bedrocks are fundamental to the story of

Strangford Lough, since they underlie much of County Down. Although Ordovician outcrops are present at a few isolated patches, the Silurian rocks mostly provide the foundation for the Lough's landscape. In many places their beds can be seen, like a tilted multi-layered cake of alternating sediment types. They are most visible in the southern half of the Lough, where they have not been covered by younger rocks and deposits. On the eastern shore they break through these later deposits just south of Kircubbin, forming a series of headlands projecting into the Lough. Rowreagh Point, Black Neb, parts of Horse Island, the Craiglee Rocks and Ballyhenry Island are all good examples. There are innumerable smaller sites, the most notable from the yachtsman's point of view being the Walter Rock immediately north of Portaferry, and the Angus Rock at the entrance of the lough. On the western shore they are less in evidence, although they just emerge at Dodd's Island near the East Down Yacht Club, Shamrock Island at Ballymorran Bay; one of the most northern is a large protrusion into the middle of the dinghy park at the Strangford Yacht Club in Whiterock. From Audley's Point south along the Narrows, they are a dominant feature of the landscape, giving a beautiful but rugged appearance to the Lough's entrance.

A wander round Ballyhenry Island, Killard Point, or a similar locality gives a good idea of the nature of these rocks. Layers of greywackes, shales, and slates, once deposited as horizontal beds of sediment, are now bent and fractured, with many of the beds tipped into an almost vertical position, folds occurring on a roughly N.E./S.W. axis. Veins of white quartz pattern the rocks, their presence due to the solution of silica from the rock under pressure, and its subsequent recrystalisation in open fractures.

In places, for example Audley's Point and Ballyhenry Island, it is possible to find the layers or dykes where molten magma was injected through fissures and lines of weakness in the rock, and subsequently cooled. They appear as layers of a compressed, crumbly pale rock running through the grits and shales. However, because these run parallel to the folds in the rock and are often covered by luxuriant weed growth, they are difficult to find. Their consistent direction reveals the way the pressures relaxed after this mountain building period, allowing tension joints to open up, through which the molten rock flowed. The earliest dykes, in the extreme south of the Lough, were subject to further periods of compression and are relatively narrow, often under a metre wide whilst the later ones, a short way further north are less compressed and may be over three metres thick.

At some stage during this upheaval, the story of Strangford Lough itself began. In addition to the extensive folding and fracturing, a wide depression formed in the bedrocks, which can be called the "Strangford Valley". Its southern end is defined by rock now forming the hills of Lecale: Slieve Patrick, Slievenagriddle, and Castlemahon Mountain, extending across to the 'mountain' at Ballywhite to the north of Portaferry. Its northern end is marked by the Craigantlet ridge, and it is at its deepest somewhere beneath Scrabo Hill. To the west lies an area of higher ground in the vicinity of Killinchy, and to the east the outcrops that act as a foundation for today's Ards Peninsula and its outlying islands. It is in this broad valley that today's Strangford Lough lies.

It is not clear how the Narrows, that deep connection with the Irish Sea, was formed. Its direction runs almost at right angles to the folds in the Silurian bedrock, and cuts straight across the high ground of Lecale and the southern Ards. An explanation may lie with the disruptive events of this mountain building era. It is possible that this is another fault created by the stresses of the Caledonian movements. Some support for this idea may come from the fact that some faults associated with the Caledonian period run in approximately the same direction. It remains unproven however, but there seems little doubt that potentially at least, the energy required to produce such a feature was available in great abundance. If the fault does exist, it is likely that various types of erosion, including glacial action almost four hundred million years later, could have exploited and enlarged the fault-line features already present, to create the spectacular channel we see today.

Relative Calm

One of the most frustrating features of any study of our planet is the fact that it is extremely rare to find a completely uninterrupted series of deposits giving a full picture of the history of a particular site. Strangford Lough and its environs are no exception to this. We know that following these turbulent events, there was a period of some fifty million years (called the Devonian), but it has left virtually no traces in the form of rock or sediment deposits, although earthquakes and laval eruptions may have continued during this time. It is therefore not clear what was happening.

It is known that about 340 million years ago, shortly after the end of the Devonian period, the area was once more inundated, this time by a slowly spreading tropical sea, although it was shallow in places. Many of the earlier forms of marine life persisted, although some had declined in importance. Trilobites were still present but did not dominate the scene as they had formerly (and were soon to die out altogether) but brachiopods, were particularly numerous. Cephalopods (literally 'head-feet'), relatives of today's octopus and squid, were also common,

ROCK TYPES FOUND IN THE STRANGFORD LOUGH AREA

PERIOD	AGE (YEARS)	DESCRIPTION
Quaternary	2 million	Blanket of glacial boulder clays, sands and gravels extending over most earlier rocks. Thinner in the south.
Tertiary	65 million	Massive complex of Dolerite sills on Scrabo Hill (also found as glacial erratics). Dykes near Newtownards and Ballyquintin Point.
Cretaceous	136 million	Not present near the Lough
Jurassic	195 million	Not present near the Lough
Triassic	225 million	A large area of Sandstone, possibly underlying much of the north end of the Lough, as well as Scrabo Hill and the Lagan Valley.
Permian	280 million	A narrow band to the north side of the Comber R., and connecting with the Lagan Valley. Matrix of coarse sands and gravels.
Carboniferous	345 million	A small remnant of limestone beds of different types lying in a depression in underlying Silurian rocks. Most notable site is Castle Espie, where it was mined.
Devonian	395 million	Not present near the Lough.
Silurian	435 million	Greywackes, Siltstones and Shales underlying all of Strangford Lough south of Ballydrain and Greyabbey. Dykes formed at end of period found in southern Ards Peninsula and Lecale.
Ordovician	500 million	Small outcrops of rocks similar to the general Silurian beds. Ards Peninsula and Ballygowan area.
Cambrian		Not present near the Lough.
Precambrian		Not present near the Lough.

How the Lough was Made 13

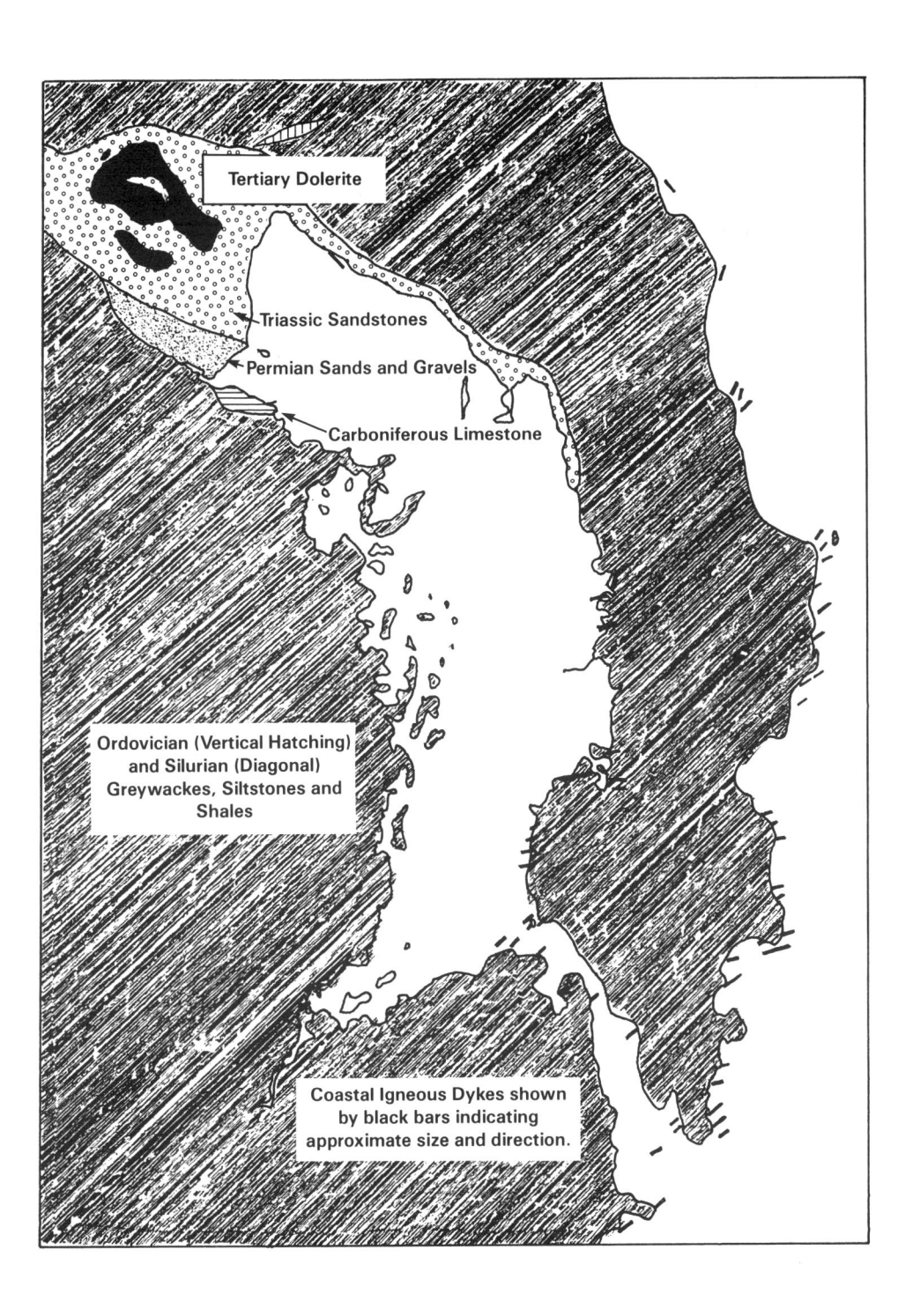

and unlike the curled nautilii still found in the Indian Ocean, had long straight shells.

New types of wildlife were also coming to the fore. Primitive sharks moved through the warm waters, and many of the lines of descent giving rise to our modern fishes were established in this period. Just as in today's Strangford Lough, much of the wealth of this marine life was due to microscopic organisms in the water - plankton. These occurred in large quantities, and amongst them were large numbers of single celled organisms encased within tiny calcareous shells. Over millions of years, these foraminifera as they are called, grew, drifted, died, and settled as fine calcareous matter on the sea bottom. Gradually this material accumulated in a thick deposit, overlying the older contorted layers of grits and mudstones. As it settled and consolidated it became a calcareous rock forming a series of limestones, in wide areas of Northern Ireland, including the Strangford Valley. Elsewhere in Britain, and to a limited extent in Ireland, swampy terrestrial habitats supported dense forests of giant tree ferns and club mosses. Their remains, preserved through incomplete decay, and subsequently compressed, produced the black rock we now call coal; as a result this phase is referred to as the Carboniferous era.

There is now little Carboniferous limestone to be seen in place around the Lough or elsewhere in County Down. It is unlikely to have formed a thick deposit, and subsequent uplifting, with exposure to aerial erosion has removed virtually all traces.

There are two known exceptions in Strangford Lough, however. The most important of these is to be found at Castle Espie, near Comber. Here a small area of limestone still remains (possibly extending some way under the north of the Lough), protected by its enclosure in a depression of the underlying Silurian rocks. In the nineteenth century the site was quarried for agricultural lime and building stone, whilst the overlying clay was used for bricks and pottery production. These excavations created a series of deep pits which are now flooded, and form the basis of the Wildfowl and Wetlands Trust bird sanctuary and wildfowl collection. The fossils found during the excavations suggest that about 300 million years ago some of the wildlife of the area was quite spectacular, with one cephalopod species (*Orthoceras*) reaching two metres in length. The fragments of sea lily stalks are also very numerous, seen as countless small circles a few millimetres in diameter embedded in the rock. Some of the stone is most attractive, stained a deep red-purple by leaching from rocks above, and it was often used for table tops and decorative architectural features, a good example being parts of the windows and some gravestones at Tullynakil Church ruin two miles south of Castle Espie.

How the Lough was Made

The other exception is the presence of large quantities of limestone drift material found in many places round the Lough, almost certainly having been carried around by glacial movements in the last fifty thousand years. The Limestone Pladdy and Limestone Rocks near Dunnyneill Island must have been particularly rich in this to have acquired their names, but much of this material was removed for agricultural lime in the nineteenth century. However, close examination of the fine shingle along the central spit of Dunnyneill Island sometimes reveals countless fragments of the sea lilies broken out and washed away from their surrounding rock. Elsewhere lumps of the red limestone (probably most derived from the Castle Espie beds) occur on many of the beaches (dip them in water to see the fossils) along with lumps of a pale grey variety, frequently bored into by bivalves.

Strangford's Desert

At the end of the Carboniferous period, with the onset of major movements of the earth's crust in southern Britain and Europe, the Strangford Valley, along with much of Ireland and Britain, was slowly lifted up once again to form a low undulating landscape. It is likely that most of the limestone deposits were eroded as they were lifted clear of the sea and became subject to weathering. These changes heralded the beginning of the Permian period, some two hundred and eighty million years ago, which was followed by the Triassic. During both these periods (collectively called the Permo-Triassic) the climate was almost arid, and desert or semi-desert conditions with a sparse vegetation prevailed over a wide area. This must have been much like present day Utah in the western U.S.A., with flat desert dotted with shallow stagnant lagoons, seasonal lakes and fluctuating rivers. Dry dust and sands drifted about in the hot winds, and settled in the brackish waters. Just as in the modern desert conditions of central Australia, occasional torrents of rain produced localised flooding, followed by long periods of drought in a scorching sun.

Relics of these inhospitable conditions can still be seen around the Lough. During the earlier stages in the Permian, a deep valley, at times possibly still below sea level, extended from the north end of the Lough to Knock. This was gradually filled in by angular fragments of eroded Silurian and Carboniferous material in a sandy matrix, probably blown in from adjacent land. These deposits, or breccia, are still present in the Comber area. Elsewhere, deposits of the desert sands, sometimes red from the oxidization of iron on the sand particles, are now consolidated to form a sandstone (known by geologists as Sherwood, formerly called Bunter). This has been used in a number of substantial buildings in the

Dried mudcracks over 200 million years old; a boulder of Triassic sandstone on Dunnyneill Island.

area, for example the Newtownards town hall, and the Robinson and Cleaver building in Belfast. Careful examination will reveal the various layers of sands laid down within the rock. At one time these deposits must have filled the whole of the Strangford Valley, forming a deep layer of horizontal sandstones stratified according to the variations in the arid conditions that took place over their 30 million year development.

Since the Triassic period, as we shall see later, considerable erosion took place over the whole area. As a result the only substantial remnants of these rocks now occur in the deeper parts of the Strangford Valley between Newtownards and Dundonald, and best seen on Scrabo Hill, where they are at their thickest and have been quarried. Similar rocks also occur along the northeast shore of the lough nearly as far as Greyabbey. They can easily be seen from the Portaferry Road just south of Newtownards, as a series of lowlying outcrops of red rock on the shore near the road. Every time there is a south or southwest gale during high tide more of the sand particles are released once again to give today's sands in that part of the lough their distinctive reddish tint.

The remnants of the ancient desert still reveal traces of the bleak conditions of their creation. Boulders, when split open, reveal surfaces of the desert floor exposed for the first time in millions of years, and curiously, it is often the very brief phases of wet weather that have left their traces in the rock. Surfaces of cracked dried mud provide evidence of flood and dessication. The desert was not devoid of wildlife during this time; small animals left tracks across the mud, presumably wriggling between shallow pools as they dried out; a small dinosaur-like reptile, probably *Ticinosuchus*, left its footprints as it wandered between waterholes in the arid landscape around Scrabo.

Volcanos, Dykes And Sills

The desert conditions of the Triassic period were followed by a return of the sea over much of the British Isles. These conditions lasted over one hundred million years, encompassing both the Jurassic and Cretaceous periods. In Antrim, deposits from the latter period have resulted in the spectacular chalk cliffs embedded with flints that rise beside the coast road north of Larne. However, in the Strangford Valley, such deposits, if they occurred, were probably a thin skim of material and were quickly eroded, so that no traces of this time remain. We do find pieces of chalk and flints on the shores, but these have been carried in by the glaciers of more recent periods.

About 65 million years ago, at the start of what is now referred to as the Tertiary period, the part of the earth's crust now occupied by Europe was once again disrupted. Massive upheaval, volcanic activity, and mountain building, was caused by further shifts in the positions of other continental blocks that ultimately affected areas as far apart as India and America, creating the Himalayas, Alps, and the Andes. Northern Ireland, including County Down and the Strangford Valley, although on the periphery to some extent, did not escape the effects of these violent events.

The first stage of this process was a series of continental movements that lifted up the land masses of northern Europe and the British Isles, so that Ireland, Britain and Europe became one block. For the Strangford Valley, filled with deposits from the earlier desert conditions, the first direct effect from the gradual uplifting of the whole land mass was an increase in the rate of erosion. This resulted in the loss of the thin Cretaceous layer (if it existed) and the upper, and most recent deposits of the desert sandstone.

It was only the beginning. Soon, forces deep within the earth's mantle began to develop to a strength not seen for several hundred million years. In the area now occupied by Northern Ireland these forces resulted in various features, including vertical fractures running on a NNW-SSE line, up which molten lava flowed breaking through to the surface in a variety of ways. In other places a number of massive and violent volcanoes developed, whose remains can still be seen in all the northern counties of Ireland. Examples of Tertiary volcanoes in Antrim are at Carrick-a-Rede, now with its famous rope bridge, and at Slemish. Elsewhere, eruptions produced quantities of molten lava, gases and superheated steam, much of which spread out in a series of lava lakes and flows from long fissures. (To the north, the flows ponded, and their subsequent slow cooling formed the Giants Causeway). A succession of sheets of lava spread out towards the Lagan valley, and extended further south, but the lava flows are most conspicuous on the north side of Belfast Lough where they cooled to form the black basalt rocks whose later erosion produced the sharp outcrop of Cave Hill, with the Antrim Plateau behind it.

On occasions molten material, driven by hot gases, forced its way between cracks and fissures in the older beds, which often were scorched, becoming pale and brittle. Several horizontal injections, or sills, penetrated the beds of Triassic sandstones in various places, including the Scrabo area. As these sills, cooled, they formed very hard dolerite (a rock slightly coarser in consistency than basalt) appearing as pronounced dark layers in the beds of desert sandstone at Scrabo.

Later, throughout the country, the combined effects of wind, rain and rivers cut through the layers of rock, sculpting the valleys and glens that form the landscapes of Ireland. Erosion of the sandstones in the Strangford Valley continued, and with the exception of an area along the N.E. shore of the Lough (and possibly under this end of the Lough as well), all of this material was removed, and with its departure, the underlying slates, grits and shales began to re-emerge.

Scrabo Quarry, Newtownards, with its Tertiary dolerite sills ('d') in the Triassic sandstone. A volcanic vent ('v') is in the extreme left.

Thus the Valley slowly began to assume some of the contours and elements that feature in today's landscape. In the far north of the Lough protected by its massive dolerite sills, the sandstone survived, gradually being exposed to form the spectacular promontary that we now call Scrabo Hill, dominating the lowlying, heavily eroded landscape around it. Its height above its surroundings is considerable, but in reality the original thickness of the ancient desert deposits was probably much greater, giving some idea of the long time it took them to accumulate.

During this period, some thirty-five kilometres to the south west, molten material was also being injected into the Silurian rocks, this time welling up to form enormous blisters of magma which slowly cooled to produce the attractive crystaline rock we now call granite. Subsequent erosion stripped away the upper skin of Silurian rocks, leaving the rounded outcrops of the Mountains of Mourne, and today they still influence the more distant elements of the Lough's landscape.

The Tertiary period, by contrast with our own apparently stable era, would have been both frightening and spectacular to our eyes. It would also have been extremely disruptive to the plant and animal life, and particularly so in the case of the lava flows which are known to have trapped the charred remains of vegetation as they flowed over the ground. In spite of this, there must have been long periods of relative quiescence between phases of turmoil, when wildlife was quite abundant, to judge by some of the material preserved from other sites. The climate was gradually becoming cooler and moister, in keeping with "Ireland's" continuing drift northwards, and relatively rich soils may have developed. With conditions possibly somewhat akin to those found today in southern Europe and north Africa, it is possible that plants and animals still familiar in those parts of the world, like Fig trees, Cinnamon, palms, Magnolia, Sequoia and Oak may have flourished in County Down. Grasses were now increasing considerably in importance, particularly towards the end of the period. Birds and mammals, having been released from competition with the large reptiles which had died out by this stage, developed rapidly. It is possible to speculate that throughout the world this was a time when the forebears of many of today's animal types were getting established, and taking over the various roles formerly dominated by the dinosaurs and their relatives.

With continuing erosion, and the excavation of the Strangford Valley from the layers of sandstone laid above it, large areas would have become lowlying. At the same time, global sea level rises and further contortions of the earth's crust separated Ireland once again from Britain, and flooded many of the inlets and bays around its coast, including the Strangford Valley. Thus, by the end of this period it is possible that the

Valley may have contained a sheltered, relatively shallow area of sea, separated from the main body of the Irish Sea by a string of rocky islands and reefs in the area where the Ards Peninsular now lies. Some of these outcrops are still with us in the shape of the Copeland Islands, Burial Island, and the Portavogie group. The remainder now lie under later deposits that we will be examining in a few pages.

The marine life at the start of this period was probably characterised by animal and plant types found today in many warm temperate seas of the world, but there is little information on its precise nature from deposits about the Lough. Towards the end of this long period, progressive cooling would have had a major influence, as may be inferred from other sites in Britain. The first of these is that both evolution and extinction of many marine life forms was continuing rapidly, reflecting what was occurring on land; fewer than five percent of the marine animal types recorded as present at the start of the Tertiary period are still alive today, whilst of those present at the end, about half can still be found.

The second major feature is that changes in sea temperatures apparently brought about a redistribution of many species. This aspect has to be treated with caution, since in many cases we are still unclear about the precise distribution of many living forms. However, with this caveat in mind, it is worth noting that many late Tertiary sites in Britain show a gradual shift from levels dominated by fossils of animals characterised by a southern distribution, like types of *Nautilus* for example, to more recent layers holding temperate, and even arctic forms like the little bivalve *Yoldia*. Such findings strongly suggest a gradual southwards shift of marine faunas in a time of cooling sea temperatures.

Snow and Ice

Nothing, not even ancient structures such as rocks, remains precisely the same. No sooner had the array of geological structures described above been laid down in much the form we can see today, and the rather fertile and attractive regime of Tertiary wildlife become established, than a new regime slowly began to exert its influence.

So far, the climate experienced by that part of the earth's crust we now call Ireland had varied considerably, generally in accordance with its slow but inexorable drift northwards over the surface of the planet. Now, the world's climate itself, which during most of this long history had been relatively warm, began to cool. The first signs were further southward shifts in the distribution of many of the temperate life forms, as they responded to the increasing frequency of cold winters, frosts, and later, snow. A slight reduction in sea levels, due to greater volumes of sea water being trapped in the northern ice cap, resulted in more arid conditions.

How the Lough was Made

This, together with the decline in the woodlands, produced cool open prairie grasslands attracting large herds of herbivores; the deer, horses, bison, and possibly including Aurochs (*Bos primigenius*), the wild ancestors of our modern cattle.

By about two million years ago, the climate had cooled sufficiently for the first permanent ice caps to be formed on the highest terrain in Scotland, the Alps, and Scandinavia. The Great Ice Age had begun.

The climate continued to chill even further, and these ice caps, formerly localised in the coldest areas, began to spread. Occupying the valleys and glens that had been excavated by the rivers and streams, glaciers slowly edged down from the upper valleys onto the lower ground. Eventually they merged and accumulated to form massive sheets of ice. These were probably well over a thousand metres thick, to judge by those still existing today near the poles. Glaciers from Scotland spread across Antrim and the Lagan valley, extending southwards across north Down towards the Strangford Valley. Cracked and contorted by internal pressures and strains, often stained by silt and other materials trapped within the ice, and studded by quantities of gravel and boulders, they slowly rasped their way over the Strangford Valley. Below the glaciers, hillsides were scoured, narrow valleys opened out, and areas of bedrock were scored and scratched.

It is important to remember that the Great Ice Age was not one continuous period of unremitting cold. Lasting nearly two million years, there were phases of warmth in which the glaciers retreated, and collectively these may have been of a greater duration than the arctic conditions themselves. During these warm spells, the climate became quite mild, and on occasions, apparently even warmer than today's climate. It has been speculated that the Ice Age may still be continuing at present, and that we are merely experiencing one of these periods of interglacial warmth (temporarily enhanced, perhaps, by the "greenhouse effect" caused by the release of carbon dioxide from fossil fuels). However, it may be several thousand years before we know for sure whether we should expect a return of the glaciers!

For much of the time, the ice that scoured the landscapes of Counties Antrim and Down originated in Scotland. However, a relatively late development was an accumulation of ice over the basin now occupied by Lough Neagh. This gave rise to an outwelling of ice which moved across County Down in approximately a south easterly direction. This has left the Lough with most of the glacial features we see today, their general orientation reflecting this direction of flow. One of these is the characteristic profile of Scrabo Hill. As the ice scoured its way over the lowlying landscape of northern Down, it flowed over and around the outcrop of

Scrabo where the sandstones were protected by the sills of hard dolerite. Although the east facing slope has been extensively modified by quarrying, the basic shape still reveals the effects of the glacier. On the northwest side of the hill it is rough and craggy where the ice attacked the hard dolerite cap. To the southeast it tails off into a long stream of debris embedded in clay and moulded by the flow of ice, now supporting prosperous farms on a rich glacial till extending through the townlands of Killynether, Castleaverry and Ballyrickard. The crag-and-tail form of Scrabo Hill can be seen from many spots about the north of the Lough, including the Maltings, Island Hill, and all the way along the Newtownards - Comber road.

The most distinctive glacial features of the Strangford Lough area are the drumlins. These numerous, small rounded hills, mostly of boulder clay, but often with rock cores, give rise to the popularly named 'basket of eggs' scenery. They can be found throughout the length of the Lough, but at the southern end there is little more than a thin skim of glacial deposits and the drumlins, though present, are not particularly well developed, so that the Silurian bedrocks dominate the landscape. On Killard Point however, extending south towards Ardglass, there is a broad band of material (moraine) formed about 16,000 years ago as sands, gravels and boulders were deposited in the sea around the margin of a decaying ice sheet.

In the central regions of the Lough, especially north of Killyleagh towards Lisbane, Ardmillan and Island Mahee, drumlins over fifty metres high provide a varied undulating landscape. They play a major role in the wildlife and its habitats, contributing bays and headlands, islands and reefs, whilst most of the sediments to be found on the shores and bed of the Lough are derived from the boulder clay.

The formation of drumlins is still a matter for lively debate, and at first sight it does seem curious that these small hills, composed primarily of clay, should have survived the onslaught of these massive glaciers. In reality, boulder clays were highly compacted by the weight of ice bearing down from above. The generally held view is that some lower sections of the glaciers, particularly in old ice near their edges, became heavily overloaded with silts, gravels, boulders and other debris. This rasped over the ground surface beneath the glacier, and in places the friction was particularly intense, possibly where protrusions of harder bedrock still persisted. In such conditions it is possible that large areas of ice and debris splintered off from the main glacier. Friction between these sections and the main flow, possibly developing a type of wave motion, then shaped the deposits of debris, streamlining them in the general direction of ice movement.

How the Lough was Made 23

Drumlins interspersed with marshy hollows on the west of the Lough

A drive down the Ballymorran road on the western side of the Lough gives excellent views of this rare and curious scenery. The highest drumlins are about a kilometre from the shore on the upper edge of the Strangford Valley. These grade into lower drumlins forming headlands with intricate bays between, and finally, as our gaze turns towards the lower parts of the Strangford Valley, we find the open waters of the Lough, with a scatter of partly submerged drumlins forming islands and pladdies, or reefs.

On a smaller scale, as the glaciers eventually melted away in their latest retreat, some 12,000 years ago, they left many other boulders, stones and silts as souvenirs of their presence, deposited more or less as they tumbled out of the melting ice. Often these materials, erratics, as they are called, had been carried for some distance, and frequently from areas characterised by very different rocks and geological formations. Thus to the southeast of Newtownards by the Portaferry road lies the Butterlump Stone, a massive boulder probably weighing over 130 tons, literally dumped on the shore. Again, to the south of Greyabbey there is the Wren's Egg, even larger, and there are equally startling erratics in many other places in the Lough. These are composed of dolerite, and have been probably carried down from Scrabo by the glaciers, and deposited when the ice melted. So, if you climb down to the shore, onto the

The Butterlump Stone. 130 tons of dolerite, probably carried from the Scrabo area by glaciers, now resting on outcrops of Triassic sandstone.

sandstone rocks at the Butterlump Stone, you can stand on the remains of a desert that existed some 225 million years ago, examine a piece of volcanic material thrust up 65 million years ago, and transported in an ice sheet that melted a mere 12,000 years ago!

Other rocks were also shifted about and now lie as stones and boulders amongst the countless pieces of shale. The remnant patch of limestone at Castle Espie was scraped and a stream of limestone fragments carried southwards down the Lough, where they can still be found. Chunks of sandstone were also carried along and these can be found on many of the islands and shores. Some material was brought in from further afield, for example pieces of chalk and flint from Antrim, various types of granites and granite derivatives, including the distinctive fine grained greyish granite from the lonely outcrop of Ailsa Craig, off the Ayreshire coast.

Wildlife during much of this bleak glacial period of Strangford's history would have been more akin to that of today's arctic. Apart from a few lichens, sedges and mosses growing on any outcrops possibly left uncovered by ice in the Mournes area, like Slieve Bearnagh and Slieve Binion, there would have been virtually no vegetation. Polar bears (*Thalarctos maritimus*), Snowy Owls (*Nyctea scandiaca*) and Arctic Foxes (*Alopex lagopus*) may have been present on the ice.

The various stages of the warmer interglacial periods allowed the wildlife of the time to develop more fully, and a number of deposits found in Ireland give some idea of the species present. With each retreat of the glacial conditions, vegetation developed and colonized the new ground. Initially this was probably of low growing tundra herbs, heathers, sedges and Dwarf Willow (*Salix herbacea*). Wildfowl and waders would have nested much in the same way that many of the Lough's wintering birds today move north to the tundra regions of Europe and North America to breed. As we shall see later, their use of such areas, and particularly the ice-free refuges during the glaciations, have profoundly influenced the migration patterns of the birds that overwinter in the Lough and elsewhere today.

The continuing improvement of the climate between glaciations eventually would have allowed pioneering shrubs like Sea Buckthorn (*Hippophae*) to become established. These would have been followed by the development of extensive forest, dominated by Oak (*Quercus*) with Pine (*Pinus*), and later, Yew (*Taxus*) and Rhododendron (*Rhododendron ponticum*). Finally, as temperatures dropped towards the end of the interglacial period there would have been a return to tundra conditions before the next incursion of the glaciers.

The Latest Thaw

About 12,000 years ago the climate gradually started to become warmer in the progression towards today's conditions. This probably occurred in fits and starts, and the amelioration in the arctic conditions was patchy, with some upland areas of ice still lingering long after most lowland areas were clear. Scouring by the glaciers had left a terrain of bare rock, boulders, barren drumlins of boulder clay. A network of icy streams supplied by glacial melt waters and persistent snow, sleet and rain, trickled through a maze of finer glacial deposits, to enter a cold, dark sea.

The events that followed the latest retreat of the glaciers provided the final touches to the Lough's topography and wildlife, lying within the ancient Strangford Valley. The most fundamental of these was the chain of effects caused by the melting of the ice sheets. This released a vast quantity of water back into the world's seas. However, the landmass also rose in response to the removal of the burden of ice, and together the two processes triggered a complex series of adjustments in the land-sea relationship. In County Down and much of north east Ireland the sea level was initially lower in relation to the land, and then some eight thousand years ago began to rise, eventually to as much as eight metres above today's level.

The higher sea levels of those times left their mark on the scenery of the Lough, in particular on the shape of the drumlins. Wave action,

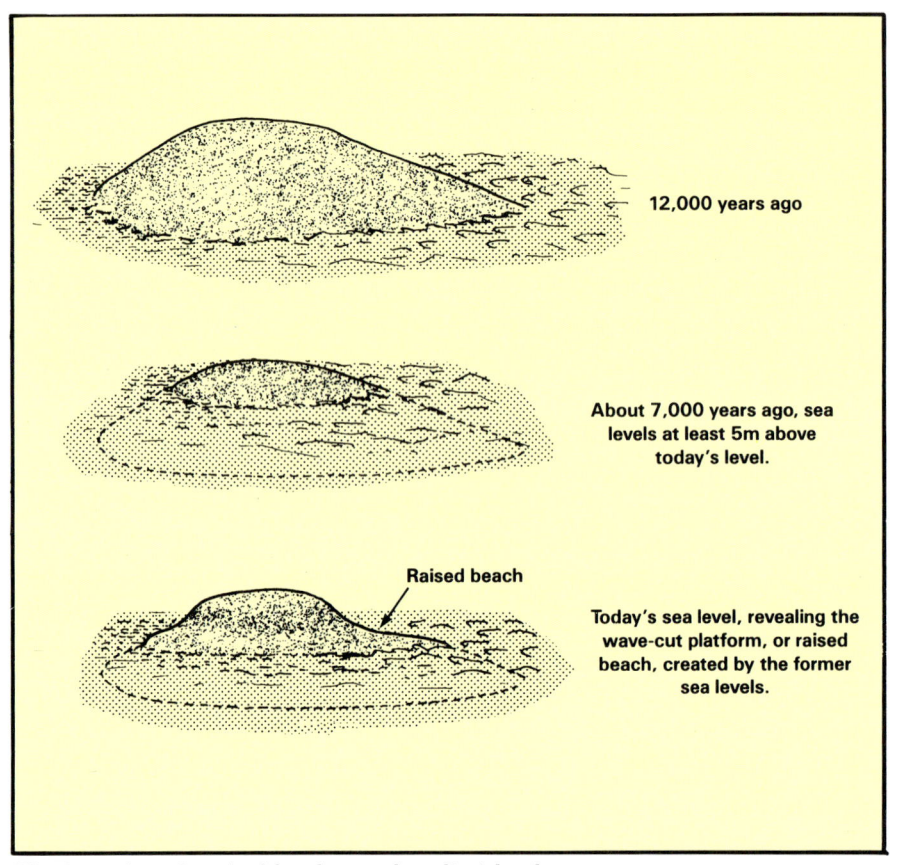

[Diagram showing three stages: 12,000 years ago; About 7,000 years ago, sea levels at least 5m above today's level; Today's sea level, revealing the wave-cut platform, or raised beach, created by the former sea levels. Middle stage shows "Raised beach" labelled.]

The formation of a raised beach on a drumlin island.

between four and eight metres above today's high water level, cut a distinctive notch in the profile of many of the drumlins, and underlying each of these there is a platform gently sloping towards the existing shore. This is the old beach, usually called 'raised beach,' and below the soil and vegetation, it is still possible to find wave-tossed pebbles, marine clays and sea shells. The prevailing wind action is from the south west, and almost certainly was then, so the west and south west faces of the drumlins show the greatest signs of erosion with the steepest wave-cut banks and often the most extensive raised beaches. Much of the road from Newtownards to Portaferry is on, or near, the raised beaches, with a steep bank to the left of the road. Out in the main body of the Lough, many islands show this feature clearly; Jackdaw Island in the south of the Lough is an obvious example, whilst the distinctive notched profile of Chapel Island off Greyabbey can easily be seen from the road near Mount Stewart.

How the Lough was Made

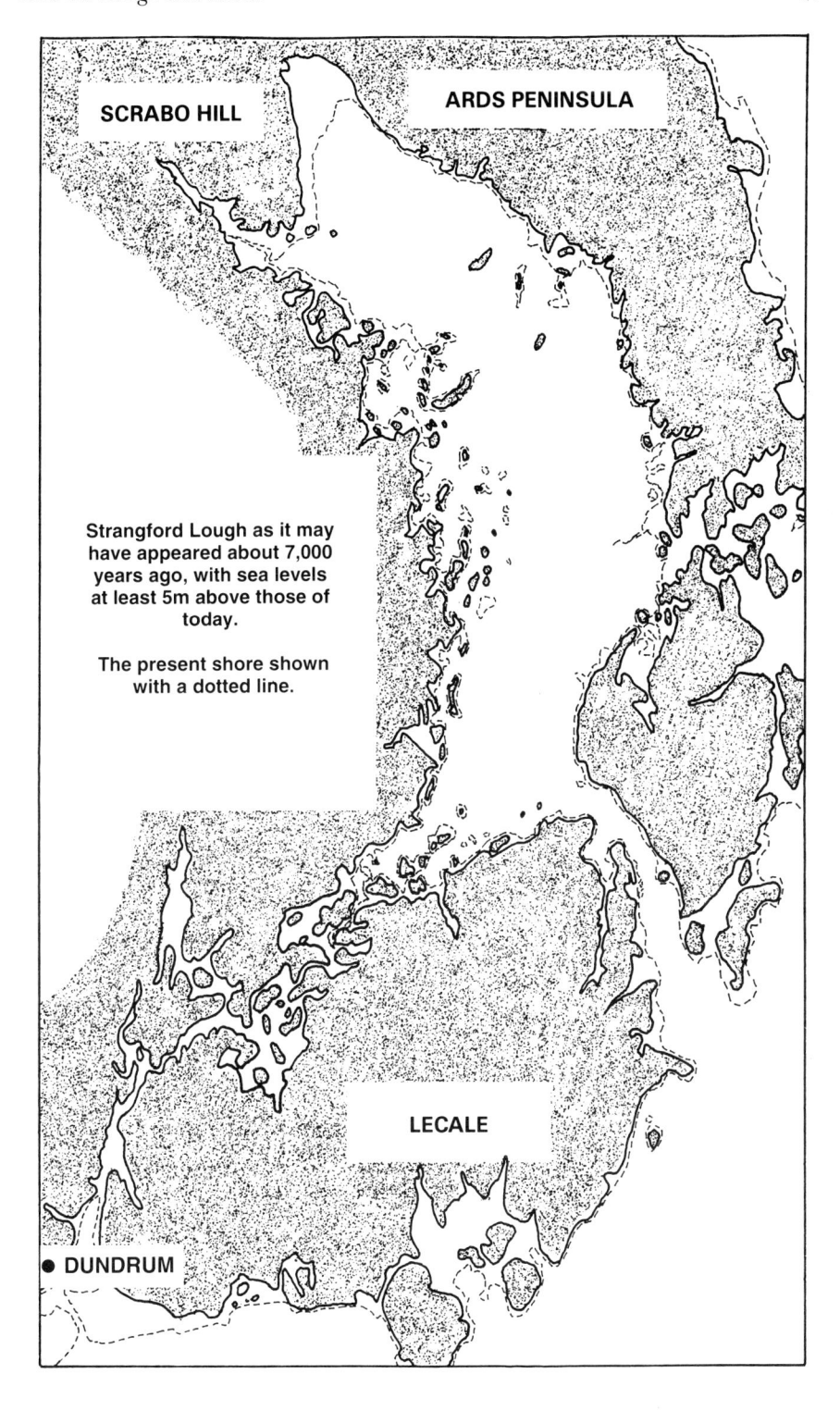

SCRABO HILL

ARDS PENINSULA

Strangford Lough as it may have appeared about 7,000 years ago, with sea levels at least 5m above those of today.

The present shore shown with a dotted line.

LECALE

● DUNDRUM

There was also a significant effect on the shape of the Lough and its contact with the Irish Sea. Examination of the contours of an Ordnance Survey map reveals a number of places where the land is particularly lowlying, and often barely ten metres above today's sea level. Although at present one can only speculate, it is possible to infer a major connection between the Lough and the Irish Sea across the Ards Peninsular in the valley of the Blackstaff River and the Kirkistown area. Further south, the hills of Silurian bedrock at Ballywhite would have been an island. The Lough would also have had an outlet through to Dundrum by way of the Quoile valley, isolating Lecale. Marine sands of this period have been found at Woodgrange, inland from Downpatrick; in mediaeval times, by virtue of the tidal Downpatrick marshes, the area was known as Isle Lecale. With these extra connections between the Lough and the open sea, the patterns of tidal currents and sediment distribution may have been different to today's regime, and as we shall see later, could possibly have left their influence on some of today's marine life.

Another major process following the retreat of the glaciers was the series of changes in the climate, as northern Europe emerged from the rigours of the Ice Age. Associated with these were changes in the wildlife of the Lough and its hinterland, and the process was similar to that of the earlier interglacial periods. Initially, with arctic conditions still prevailing, these were much as today's tundra regions in northern Norway. Open country was dominated by tracts of Dwarf Willow, Birch scrub (*Betula nana*) and other arctic plants, interspersed with icy lakes. The Mammoth (*Mammuthus primigenius*) failed to survive the last glaciation, but the giant Irish Deer (*Megaceros giganteus*) must have found an ice free refuge in the south or west during the last glacial advance, and was able to return northwards after the retreat of the glaciers. Its remains have been found near Downpatrick and in the marshes, at Scrabo, and several places near Portaferry, and it must have been a spectacular animal; as big as a large horse, with antlers up to 3 metres from tip to tip. Reindeer (*Rangifer tarandus*), Hyena (*Crocuta crocuta*), Brown Bears (*Ursus arctos*), Lemmings (*Lemmus lemmus*), Wolves (*Canis lupus* - finally exterminated in the 18th century) and Arctic Foxes returned northwards and roamed the Irish postglacial scene, many of these now only found in the tundra and arctic.

In the Lough similar changes were taking place in the marine life. Initial colonizers were species like the bivalves *Astarte borealis* and *Pecten groenlandia*, species that, as their names suggest, have a pronounced northern distribution. Later, with further warming of the climate, most of these disappeared or declined, as they shifted further north. At the same time, warmer-water species like the Carpet-shell (*Venerupis decussata*)

The River Quoile. This valley may once have connected the Lough with Dundrum Bay.

made their appearance. The northern aspects of the Lough's wildlife may not have completely disappeared, however. Certainly some northern species can be found in the Lough today, and these could have persisted as relics of those early post glacial times, able to survive because of the Lough's equable temperature regime. A good example is the northern starfish *Leptasterias mulleri*, whilst some animal communities, like that based on the Horse Mussel (*Modiolus modiolus*) have features that do not seem to be encountered nearer than Scandinavia.

The post glacial rise in sea level eventually completed the final separation of Ireland from Britain by 5000BC. However, before this process was complete, a significant number of species (possibly even man) used the land bridge, thought to have linked the north coast of Ulster with Islay or the Mull of Kintyre (some feel other links may have been much further south). The selection of species which were able to cross over in the ameliorating conditions before the connection was finally severed, have largely determined the composition of today's Irish terrestrial fauna and flora.

Many species common in Britain never crossed the land bridge. To this day, Ireland has little more than half the number of plant species; notable animal absentees include snakes, Moles (*Talpa europaea*), and Field Voles (*Microtus agrestis*). Ireland has only twenty-eight wild mammals compared with Britain's fifty-five. At an early stage, butterflies like the

Large Heath (*Coenonympha tullia*) and the Small Mountain Ringlet (*Erebia epiphron* - not seen in Ireland since the 1920's) became established, to be followed by many of the non-migratory species that we find in Ireland and about Strangford Lough today. Other butterfly species that are present in England may never have reached Ireland, so that the Irish butterfly fauna is also much smaller than that across the water.

As the climate warmed, a succession of plants demanding more temperate conditions were able to spead northwards, together with their associated habitats. By about 7000BC hazel and pine woodland developed. By about 3000BC, with temperatures exceeding today's values, the climate and the vegetation took on a more oceanic aspect, with oak and alder now dominating the lowland scene, and pine and birch able to thrive some 300 metres above their current altitude limits. About 500BC there was a marked deterioration in the climate, with a distinctive oceanic tendency of mild wet summers and cool wet winters. The increase in moisture and reduced temperatures resulted in a decline in the tree cover, particularly in the higher and more exposed areas of Ireland. These conditions were ideal for the growth of *Sphagnum* moss, one of the main constituents of peat, which accumulates through the incomplete decay of vegetation in moist conditions lacking the supply of oxygen essential for bacterial growth.

On lower, more sheltered, and better drained ground, conditions did not become so severe. Round Strangford Lough, dense forests of Oak (*Quercus petraea*) and Elm (*Ulmus procera*) thrived, scattered with Ash (*Fraxinus excelsior*), and with Willow (*Salix spp.*), Hazel (*Corylus arellana*) and Alder (*Alnus glutinosa*) in the damper areas. It was only in some interdrumlin hollows and badly drained areas of bedrock, for example Inishargy, that peat bogs could develop. The wildlife of the Lough itself, now beginning to experience the early effects of man's activities, would by this time have been quite similar to that described in the following chapters.

Today's Lough

Some five hundred million years now separate the Ordovician-Silurian deep ocean and its early life forms with today's Strangford Lough. Since those rocks were laid down, a series of cataclysmic events have broken upon the scene, and long periods have come and gone, leaving thousands of metres of younger sediments which in their turn, have been largely eroded away. Although the Lough occupies the centre stage of this book, it is important to remember that its history has been subject to events whose centres have often been far away from the Lough itself, that have shaped the continents and oceans of the world as we know it.

Fate has been highly selective in determining the role the various formations play in the landscape of the Lough. The Carboniferous Limestones, and the Triassic desert Sandstones influence the quality of the glacial soils in their presence as drift material, but otherwise play only a small part today in the story of the Lough's wildlife. Their softness and friability has ensured that once lifted to any extent, the erosive forces of nature would soon return them to sediment form once again.

Instead, the most ancient periods of the Strangford Valley combine with the most recent to provide us with the unique Lough we can visit today and which holds so much wildlife. The greywacke grits and shales that were laid down before the Valley itself was formed, now provide a foundation for the Lough and its hinterland, the hills in the south, and frame the rugged Narrows with its powerful currents that link the Lough to the Irish sea. The rest of the Lough is largely overlain by deposits of the Great Ice Age, which in geological terms happened scarcely yesterday, and is probably still with us. It provides us with Strangford's tortuous shoreline, its rounded hills and islands, most of its navigational hazards, and most of the richly inhabited sediments lying on the floor and shores of the Lough. However, we do well to remember that these sands and muds also had a history before they were ground into tiny particles.

The processes that have shaped the Lough over its history continue to operate with the same intensity as they have done throughout many earlier times. The hard grits and shales are still being slowly ground away by the actions of wind, wave, rivers and streams, and the softer glacial deposits, also subject to the same treatment, are probably disappearing very rapidly indeed, when viewed on a geological timescale. It is likely that in the wider scheme of events their appearance as a feature of the landscape of the Strangford Valley will be short lived. In the meantime however, their effect is considerable, both on the physical forces of the sea operating within the Lough, and on its wildlife.

3 Tides, Winds and Waves

IF ONLY one word could be used to describe today's Strangford Lough and its wildlife, that word would be "movement". Changing tides with powerful currents, unpredictable weather patterns, and the constant rush of waves back and forth over the shores, give the Lough its character, and profoundly influence the animals and plants. They give the wildlife a restless quality, in which changes are perceived not only in terms of seasons, miles or long geological periods, but even within a matter of minutes, or within a few yards of shore.

There are times, albeit very occasionally, when the Lough seems like a large and placid lake. On a clear calm spring morning one can stand on the rocks at Audley's Point, gaze across blue ripple-free water, and feel it is almost possible to reach out and touch Scrabo Hill far away to the north.

It is only an illusion. In the Narrows, running between Strangford and Portaferry, cut deep into the contorted Silurian bedrocks, a massive surge of water flows with every turn of the tide. A restless and uneasy combination of currents, whirlpools, upwellings and powerful back-eddies, it streams along, relentlessly tugging boats off course, carrying seals and dead branches with its flow; a powerful reminder that in spite of being so landlocked, the Lough is very much a part of the Irish Sea.

Although some parts of the Lough remain tranquil even in the foulest weather, many areas experience surprisingly rough conditions, when powerful waves hammer against the shores, driving into every inlet and rocky cleft, and throwing spray some distance inland. When these waves interact with tidal currents, the effect can be spectacular! The result is that Strangford Lough has a quite extraordinary variety of physical conditions for such a small area of sea. Not only do these vary from one part of the Lough to another, they can also change rapidly - within a matter of hours, as the tide turns.

The Rising Tide

It is best to start by looking at the global picture. Tides are caused by the gravitational pull of the moon, and to a lesser extent, the sun. As the moon orbits the earth it pulls at the land or sea immediately beneath it.

When the moon is over a particular area of sea, it draws the water very slightly towards it, causing the phenomenon we call 'high tide.' Sea areas facing at right angles to the moon's influence at this time experience 'low tide.' Although there are many anomalies around the world depending on the shape and position of the various oceans, most areas, including Strangford Lough, get two high and two low tides each day. When the sun and moon are most closely aligned (new and full moon) the influence is fractionally stronger so that for a few days we get 'spring tides' when the tidal range is somewhat larger. When the sun and moon are most out of alignment (half moon) the tidal amplitude is at its least and we call these 'neap tides.'

All this may seem rather distant and hypothetical as the rising tide ripples its way over the Lough's mudflats or slowly creeps up the wall of some jetty or pier. However, of all the factors that play a role in the nature of Strangford Lough, the role of the tides is perhaps the most fundamental in shaping the Lough's character and its wildlife. The tides influence human activities too; they impose a strict timetable on yachtsmen, divers, beachcombers, wildfowlers, oysterfarmers, and anyone else who wishes to venture below the high water mark. On one occasion the Lough can display apparently clear open stretches of water, whilst a few hours later the same view reveals a maze of rock outcrops, mud banks, reefs and channels - a delight and challenge to navigation!

As the moon approaches some 384,000 Km above the earth, the North Atlantic waters come under its influence, and begin to rise around Ireland's shores. In doing so, this tidal bulge enters the Irish Sea at its southern end via St. Georges Channel, past Wexford. At the northern end, the water flows between Rathlin Island and the Mull of Kintyre, as the tidal bulge enters the North Channel. This is the body of water that concerns Strangford Lough; it is very much part of the mild waters of the Gulf Stream flowing from the North Atlantic, although occasionally after prolonged northerly gales it may have a chill sub-arctic component swept down from the waters off north-west Scotland.

As the waters flow in through the North Channel, successive places along the Ulster coast experience high water. Local tide tables give the times for Belfast with a list of correction factors to account for the shifting pattern of tides around the coast. Larne experiences high water several minutes ahead of Belfast, and Donaghadee and Warrenpoint a short while after. Thus the rising tide works southwards along the Irish coast, finally meeting its counterpart from St. Georges Channel somewhere in the region of St. John's Point.

To see how Strangford Lough's tides fit into this picture we must look at the shape of the lough itself. There are two essential features. The first

The Narrows, looking north into the Lough. About 350 million cubic metres of water flow through with every tide.

is the Lough's connection to the Irish Sea - the 'Narrows.' Guarded by the low headlands of Silurian bedrock, Ballyquintin and Killard Points, cluttered with rocks and reefs like Pladdy Lug, Angus Rock and Rock Patrick, this rugged channel runs roughly in a north northwest direction into the main part of the Lough. It is an impressive feature - well worth looking out for on flights between Belfast and London. Alternatively, go up to the old windmill just south of Portaferry. To the south in the distance lies the shallow entrance to the Narrows ('the Bar'). Closer to hand is the wooded foreland of Bankmore Hill. Between here and Isle O' Valla on the other side of the Narrows, the channel is less than 800m wide, but it is almost 30m deep. Across lies the village of Strangford and the woodlands of Castle Ward. To the north, one can see Audley's Point and Ballyhenry Island, and the waters in this area are about 60m deep in places.

To the north of these two landmarks the Narrows opens out quite abruptly into the second feature; the main body of the Lough itself. Some 24Km long, about 8Km wide (depending on where it is measured) it represents a section of Irish Sea pooled within the ancient Strangford Valley and enclosed by the Ards Peninsula. In its southern reaches there are some spots as deep as the Narrows, part of a channel more usually between 20 - 30m deep that extends as far north as Island Reagh, well towards the north of the Lough. A secondary arm of the channel branches off near Gransha Point and runs north east towards Kircubbin. Elsewhere, the Lough is relatively shallow - generally less than 10m deep, and often less than that. The statistics of the Lough as a whole are impressive. About 150 square kilometres of sea, or (approximately!) 1650 million cubic metres of seawater at high tide are held within 240Km of contorted shoreline; there are about 120 islands, and considerably more than that in intertidal pladdies, reefs and rock outcrops. The relationship between the main part of the Lough and the Narrows determines the nature of the tides, the currents, and the way these influence the materials that make up the shores and the bed of the Lough. It is also the key to understanding why the Lough's wildlife is so rich and varied.

To see how the system operates we must return to the tide rising in the Irish Sea. As the waters creep up over the outer shores of the Ards Peninsula, the shoreline rocks of Ballyquintin Point and the Angus Rock begin to experience the effect as well. At first the change is imperceptible. Gradually however, weeds, formerly draped flaccid over the stones, begin to float and swing in the water trickling between the rocks and boulders. Soon the rocks themselves begin to disappear one by one beneath the rising water. The long stretch of the Angus Rock soon appears to contract; eventually the outcrop divides in two as the rising tide separates the lighthouse from its small neighbour, the Garter Rock.

At this stage, a vast volume of water - estimated at 350 million cubic metres - is beginning to flow through the Narrows into the expanses of the Lough a short distance to the north. This surge of water (known locally as The 'River') flows over and around rocks, round jetties, plucking at weeds and releasing them only to catch them again, streaming along in some of the fastest sea currents in northwest Europe. And as it does so, the water level in the Narrows creeps up over the shores, rock outcrops and village slipways. In places, particularly in the most constricted areas between the Gowland and Cloghy Rocks, these currents can reach speeds of over seven Knots (or about 18Km per hour - 11mph). Whirlpools caused by turbulence over obstructions and irregularities on the sea bottom, and rock outcrops along the shores, pull at any boat trying to steer a straight course. Elsewhere, upwellings thrust water up from far below. These erupt as overfalls onto the surface, creating bulges of smooth water spreading outwards over the choppy water around it. Two to three hours before or after high water during Spring tides (ie. when high water in the Lough is predicted for roughly 2 a.m. or 2 p.m.) are the best times to watch for these strange current patterns. The most spectacular, present on both ebb and flood tide, is the Routen Wheel, lying about a hundred yards off Bankmore Hill (also called Rue Point).

Tides, Winds and Waves

Apparently so named by the Vikings because its sound was regarded as similar to that of cattle snoring, it represents a permanent hazard to small boats, particularly when at its most powerful in strong winds or during spring tides. With the constant murmur of these uneasy waters and their startling patterns of movement, it is an exciting, but dangerous place to visit by boat.

For a less energetic approach to the Narrows it is easy to sit and watch the ferry, or indeed any other boat, on passage between Strangford and Portaferry, swept sideways by the current. The turbulence can easily be seen from the ferry, especially about 100 metres out from either shore as it enters the main flow. Alternatively, take a walk along the beach or the shore road to the north of Portaferry. About a quarter of a mile north of the ferry slipway lies the Walter Rock, about a hundred metres off shore. It is worth pausing a moment to view the rush of water past the outcrop - watch for Cormorants (*Phalacracorax carbo*) alighting on the water nearby and being suddenly swept along in the flow. Continuing north-

Patterns of water movement in the Lough, the greater energy being shown by the larger arrows. Left, tidal currents; right, wave action.

The Routen Wheel, south of Portaferry. Currents up to 8 Knots occur in this area.

wards, it is possible to sit and watch the water rush past the beacon at the point of Ballyhenry Island.

To the north of Ballyhenry Island the flow enters the main body of Strangford Lough, and the wide expanses allow more room for water movement. As a result the current starts to slacken off. Nonetheless where the current runs along the central channel of deeper water it still retains enough strength, particularly on a spring tide, to cause turbulence as far as Dunnyneill Island about two and a half miles north of Ballyhenry. Divers who have negotiated the channel between Dunnyneill and the Long Rock, or off the Limestone and Abbey Rocks will know the strength of the currents there. Indeed there are virtually no areas in the centre of the Lough that are completely free of these currents; even at Mahee Point 13Km to the north of Ballyhenry Island, a flow of up to two knots can sometimes be detected in the form of disturbed, rather choppy wave patterns when it pushes against a breeze.

Gradually however, as we move away from the Narrows, and away from the main thrust of the currents in the central channel, the water flow becomes much reduced. Although there are some currents between the islands and headlands in the west of the Lough, for example Ringhaddy Sound, the general rule is that the more an area is enclosed, the less the energy from water movement. Some areas may be relatively deep, like Scott's Hole (33m), off the Rathgorman Pladdies. Alternatively, areas experiencing equally low levels of energy from tidal currents may be found in the most enclosed corners of the Lough, those shallow, intimate little bays like Ballymoran and Quarterland where there are virtually no signs of those powerful tidal currents that are such a feature in the south of the lough. Indeed, were it not for the rise and fall of the tide gently creeping over these stretches of soft mud locked within their drumlin headlands, these sheltered inlets could easily masquerade as lakes.

There are however, two places, away from the normal run of tidal currents, that break all the rules of the Lough's tidal system. One links Ardmillan Bay in the northwest, to the main part of the Lough in a series of narrow channels between Mahee, Rainey and Sketrick Islands. Formerly there were also connections between Mahee Island, Island Reagh, and the mainland at Ringneill, but these were closed by causeways constructed in the nineteenth century. In a sense Ardmillan Bay acts like a small version of Strangford Lough; passing through those quiet waters locked within the drumlin islands of the western shore, we suddenly find a tidal surge almost as powerful as that in the Narrows, rushing past Ballydorn ('Dorn' is an Irish word meaning 'narrow channel') as the waters ebb and flood between the Lough and Ardmillan Bay.

The other place, also appropriately called the 'Dorn,' surely qualifies as the most unlikely coastal feature of them all. 6Km to the north of Portaferry, lies a small bay with a narrow entrance squeezed between the twin drumlins of Castle Hill and Dorn Hill. Across the narrow channel lies a bold outcrop of bedrock forming a ledge, or sill. This acts very much like a dam, holding back the waters within the bay and resulting in that rarest of phenomena - a marine waterfall, as the water rushes out over the sill. Eventually the rising tide in the main part of the Lough gradually creeps up to the level of the sill, and finally laps over. Then the outward flow over the sill quite abruptly reverses and the sea floods back into the bay again.

The Turn of the Tide

Because of the constriction of the Narrows, it takes some time for each rising tide to flow into Strangford Lough, and so high tides within the Lough lag behind those in the open sea. Even within the Narrows there is almost ninety minutes' delay between Killard Point at the entrance and Audley's Point at the northern end. Within the Lough, there is a further delay of some thirty minutes.

Another difference between the Lough and the Irish Sea is that the tidal range (difference between high and low water levels) is less than four metres compared with about five on much of the open coast. Again, it is the Narrows creating this effect by hindering the tide levels in the Lough from reaching their potential maximum, and equally preventing them from completely ebbing before the next tide outside the Lough catches up again. In reality the picture is even more complex than this, as the current continues to flow into the Lough along the edge of the Narrows in a series of back eddies, whilst in the centre, the tide has turned and the water has started to flow out. Precisely why this happens is not clear, and it is a very good illustration of how much we still have to learn about the Lough.

The ebb tide is every bit as spectacular as the flow. The currents are just as powerful, and just as turbulent. The Routen Wheel is still present, though on the ebb tide it shifts about 400m south towards the Gowland Rocks. As the currents flowing out through the Narrows gather pace, the effects of turbulence within the main body of the lough disappear. Instead, there is a powerful surge of Strangford Lough water flowing out past the Angus Rock and penetrating several miles out into the Irish Sea. This gradually curls northwards as it comes under the influence of the ebb tide currents flowing northwards along the Down coast towards the North Channel. The effect is impressive. Any time there is a good breeze coming from the southern or eastern halves of the compass it can be seen from the air in the shape of broken wave patterns - but perhaps most significantly, it has even been recorded in Landsat satellite photographs taken from several hundred Km up in space.

Just as spring high tides, perhaps two feet higher than normal, creep over saltmarshes, and up against strand lines normally left dry by the tide, so the low water levels of spring tides also present a very different picture. Vast new stretches of shore and mudflats lie bare. In places the rotting hulks of forgotten boats are briefly exposed. Elsewhere, yachts in the more confined anchorages may gently nudge the bottom with their keels, and around the islands and pladdies broad bands of shining kelp weeds extend far out into the calm water. This is the best time to go exploring - in some areas the water's edge may be almost a kilometre further out than usual. Down at those inaccessible levels the ebb tide reveals secrets of the Lough's wildlife normally available only to those who dive. But beware: the full ebb of a really good spring tide is followed by a rapid rise in water level as the tide turns and advances to spring high water. It is all too easy to get cut off from the shore as the rising waters swirl and trickle around you. In such circumstances, a bootful of water may be the very least of your problems.

Water as Habitat

It is not known how much of the water entering the Lough on a rising tide actually mixes with the water already there, and how much merely

Tides, Winds and Waves 41

Rocks at Kilclief. Outcrops like these along the shores and on the sea bed create turbulence that mixes the water thoroughly (A.J.).

leaves again without any great degree of mixing. It is clear however, that the tidal flow has major implications for the nature of the Lough's waters.

Most sea areas have some degree of layering, or stratification; in other words, there are usually brightly lit surface layers stirred by the wind and heated by the sun, one or more intermediate layers, and a zone of dim or dark water intimately linked with the sea bottom. Each may differ from the others in its temperature, salinity or chemical composition, and they may vary considerably in the degree to which they mix with each other. Once again, Strangford Lough is an exception. With each tide, the turbulence of the water in the Narrows breaks up these layers, and mixes them thoroughly. This results in even temperatures and a remarkably equable distribution of dissolved nutrients throughout most of the Lough. Because of this the Lough never gets particularly cold in winter, nor especially hot in summer, and there is an ample supply of basic nutrients to all levels throughout most of the year. The situation found in the open sea, where surface nutrients are heavily depleted by microscopic algae in early summer, does not occur as there is always a fresh supply being churned up from below. It has also been calculated that the total yearly freshwater input to the Lough from rivers and streams (mainly the Comber and Quoile Rivers) is exceeded by the amount of sea water entering over a mere three tides. Thus the salinity of the Lough as a whole is virtually identical to that of the open sea - Strangford is a true sea lough - not an estuary!

There are however, some places in the Lough where the waters have somewhat different characteristics. For example, where streams and rivers flow in there are local effects due to the run of fresh water. The best example can be seen on the approach to Gibb's Island and the Quoile Yacht Club, as the boat passes westwards from the relatively blue-green waters of the Lough into the brown, somewhat murky fresh water, that has flowed out from the sluice gates of the Quoile Barrage. After heavy rain, with the Quoile in full flood, quite a sharp line between the two waters can be seen as far out as Green Island. Elsewhere it is possible to see the same division between the waters of any farmland stream and the true sea water of the Lough. In severe winter weather these are often the corners where ice forms first, since fresh water freezes sooner than sea water.

One major area that seems to operate in a slightly different way to the rest of the Lough is the shallow area lying to the north of Greyabbey and Island Reagh. To some extent this is due to the fact that the rate of water exchange with the other areas is slightly less - possibly because of the narrowing of the Lough at Island Reagh, and the long distance from the Narrows. With the tide rising over vast stretches of open mudflats the effects of summer sun and winter frost are much greater, and as a result water temperatures in this area vary considerably. Water from the Comber River can take somewhat longer to disperse, as can the outflows of the Newtownards canal or the Ballyrickard sewerage works. The combination of all these factors increases the likelihood of layering, or stratification, in this part of the Lough; this would reduce the amount of nutrient exchange in the water with considerable implications for wildlife, and for the effects of sewerage effluents. It may be possible that the one redeeming feature preventing stratification of this area is the influence of wind and wave action which must help to stir up those shallow waters. If this is the case, the balance is indeed a delicate one.

Our Unreliable Weather

The powerful tidal currents of the Lough may be at the heart of much of its character, but they are not the only force to be reckoned with. The shallow areas of the Lough, its surface waters, and its shores are all strongly influenced by the unpredictable Northern Irish weather.

By contrast with many other places, even parts of neighbouring Scotland, Ulster's climate is gentle and generally free from extremes of temperature and rainfall. This is mainly due to its westerly position in Europe, under the influence of the relatively moist and mild air sweeping

Tides, Winds and Waves

Castle Ward Bay and Audley's Point. Snow and cold weather are rare in the Lough, but it can put stress on the Lough's wintering birds (A.J.).

in off the Atlantic. Occasionally continental air masses, hot and dry in summer and often well below freezing in winter, venture into the Province. Usually however, it is not very long before the more moderate Atlantic air masses reassert themselves, and the more normal pattern of depressions moving in from the west becomes re-established.

This cycle of depressions is a familiar one. Often the depressions pass between Ireland and Iceland or over the north of Scotland, driving fronts of relatively warm south-westerly winds and rain. As they pass, the wind usually swings towards the north-west, it becomes brighter and somewhat colder, and a succession of squally showers trundles across. Following this there is usually a phase of better weather before the next depression rolls in. This pattern dominates Northern Irish life; farming, fishing, and all outdoor social events are rarely free from the gamble of the day's weather conditions.

Strangford Lough is subject to essentially the same weather regime as the rest of the Province, but the rainfall is considerably reduced by the presence of the Mountains of Mourne some 35Km to the south-west. When the moist south-westerly air rises and cools over the mountains, the humidity reaches saturation point, and it rains. In the centre of the Mournes, the misty slopes of Slieve Bearnagh receive a thoroughly soaking 1770mm of rain each year. Areas downwind of the Mournes are progressively drier; Newcastle receives 1335mm of rain, Murlough Nature Reserve near Dundrum, 950mm, whilst Lough Cowey near the eastern shore of Strangford Lough gets 842mm, or about half the annual rainfall of the Mournes. This rain shadow as it is called, gives Lecale, the southern Ards, and south-western parts of the Lough the distinction of being one of the driest parts of Ireland.

In spite of the reduced rainfall, the south of the Lough is quite often influenced by fog and mist. Warm humid spells often cause this when the air chills over the sea surface. When moist air cools to a certain temperature (usually referred to as the 'dew point'), the water condenses into countless tiny airborn droplets. These drift in as mist, but where they come into contact with the ground, branches, or telegraph lines they can settle out as dew. Sometimes in early morning or late evening, a long bank of fog can be seen about a mile off the Angus Rock at the entrance, drifting slowly in on a light breeze. There are times when it is possible to climb to the top of Kilclief Castle and look out over the mist to see the Angus Rock lighthouse apparently hovering in mid air, and disembodied masts of yachts swinging back and forth as their skippers in the murk below negotiate the swell across the bar. In the main part of the Lough however, the presence of warmer land nearby usually lifts the air temperature above dew point, and in daytime the mist clears, almost appearing to retreat back down the Narrows again.

Another very important feature of the local climate is the influence the Lough has on adjacent land areas. Since the sea changes temperature more slowly than the land, it exercises a dampening effect on local temperature variations. In summer, land areas immediately next to the Lough can be several degrees cooler than areas further inland, and there is often a chill onshore breeze. Most importantly, the same effect prevents temperatures from dropping very low in winter, so a zone of farmland round the shores enjoys winters that are unusually free of frost and snow - on occasions the effect is quite startling, as one drives from inland County Down to the shores of the lough. This is of enormous economic importance; farmers in east Down have one of the longest growing seasons in the Province. If it were not for the constant recirculation of the Lough's waters provided by the Narrows and the powerful tidal currents, much of this effect would be lost.

Wind and Wave

The rainfall patterns and the other local weather features undoubtedly have a major effect on wildlife, but it is the wind that has the most obvious influence, particularly along its shores. This may seem strange; the Lough is completely landlocked and so it should be marvellously sheltered compared with the open Irish Sea. This is true, but its waters are sufficiently open for the wind to play a major role in determining its wave patterns.

The extent of wave action varies considerably from one part of the Lough to another. Wind blowing over an area of open water requires some distance ('fetch') to build up waves. As the wind drives in most frequently from between the south and the west, the western shores are relatively free from strong wave action as the fetch is so small. Moreover, the high drumlin headlands around Killyleagh, Killinchy and Whiterock, together with the cluster of tiny islands and reefs, all help to protect places like Ardmillan, Ballymorran and Quarterland Bays from the prevailing winds and even from the rare incursions of strong easterlies.

On the open waters, conditions can vary enormously. It is possible to be in calm water, perhaps gently rippled by a soft breeze, and then within a few minutes' boat journey to be in a choppy sea with waves seeming to attack the boat from every direction. Often this is due, not to the wind or tidal current alone, but to the combination of the two. When wind and current are running in approximately the same direction, the waters tend to be relatively calm. A few hours later, when the tide turns, and current and wind are in opposition, a fierce sea can develop with standing waves breaking against each other as they fight against the current. This effect is especially dramatic where the current runs over the shallows at the Bar off the Angus Rock and it meets the force of an autumn gale. A long line of 10m breakers marks the area, on occasions even breaking over the pillar on the Pladdy Lug. In boat passages they give the feeling that one is being sucked into a raging wall of water; white spray whipped off the wave tops drives against a dark grey sea and obscures the view of reassuring landmarks in the distance.

This type of effect is not confined to the Narrows. Off many headlands and islands where there is some current strength (for example, north of Audleystown on a rising tide in a strong westerly, the Limestone Rock, and the Long Sheelah) there may be strong wave action when the tide runs against the breeze.

On the eastern side of the Lough, the dominant south westerlies have had at least 6Km of open water in which to build up the waves. If the wind is coming from the south, the fetch may even be as much as 12 Km. All that energy dashes itself against the shores of the Ards Peninsula. From Marlfield to Kircubbin, every rock or boulder headland is swept by waves during the winter depressions, with places like Ringburr Point, Black Neb and the outer rocks of Horse Island taking the brunt. To the north of Kircubbin the effect is hardly less with headlands around Bloody Burn and Herring Bays steadily being eroded under the onslaught with a wave surge that may reach down some 10m below the surface of the water, sucking and shifting the sea bed materials close in to the shore.

As these waves swept before the wind run up against the shore, they batter the rocks and stones, forcing themselves into every nook and cranny, sometimes with explosive effect. In doing so, they carry a cargo of stone fragments, debris, and other small particles that constantly grind away the materials of the shore. Over thousands of years, the ancient Silurian rocks, and the more recent glacial boulder clays, have all been subjected to this constant weathering. As we saw in the last chapter, this process left its mark in the shape of truncated drumlins and wave-cut beaches left dry after the changes in sea level some five thousand years ago.

Today the process continues and it is no less insidious, penetrating every corner of the Lough, with even the most enclosed corners being subject to some measure of wear and tear. Human efforts to meet the challenge of the waves with causeways, tidebanks and sea walls may retard the process, but in the long term they are all subject to the same inexorable laws. The Island Hill causeway near Comber for example, although sheltering behind the bulk of Rough Island, will need constant repair, as the waves hammer and suck at the material under the concrete.

Bedrock, Boulders, Shingle and Sand

By now it should be clear that we find in Strangford Lough an extraordinary combination of tidal currents and wave patterns. The effects of

The northern mudflats. Rich in invertebrates and algae, they act as a giant filterbed, maintaining water quality, and supporting enormous numbers of wintering birds.

Tides, Winds and Waves

A February storm, with waves attacking the tide bank at Anne's Point, Greyabbey (D.A.).

these on the marine life are many and varied, but they have one fundamental factor in common: energy. Every current driven by the turn of the tide, every wave running before the wind, has its own store of energy to dispel upon the materials that make up the Lough shores and sea bed. That energy comes from the water movement; in other words, it is hydraulic energy. In the depths of the Lough, below the influence of wind and wave, hydraulic energy is mainly derived from the tidal currents. Closer to the surface the effects of wave energy gradually become stronger. Waves are not merely surface phenomena; each wave as seen on the surface represents a cyclic movement of water, and its effects, depending on the wind speed, may reach as much as 10m depth, and on the shore they are usually the dominant source of hydraulic energy.

The strength of this energy, from whatever source, fundamentally influences the nature of the materials in any locality. Small, light particles of silt or fine sand, are easily shifted and so they only tend to settle down in areas of very low energy. At the opposite end of the spectrum, it clearly takes a massive amount of energy to shift boulders or even outcrops of bedrock. The result of all this is that shore or seabed areas experiencing high levels of hydraulic energy derived from wave or current action (or both together) tend to have coarser materials, whilst areas away from tidal currents and in sheltered localities receiving low levels of energy are therefore characterized by very fine sediments.

This process rapidly becomes clear when we look at the bed of Strangford Lough and its shores. On the bed of the Narrows, nothing is allowed to rest for long. Small particles, stones and even boulders can be moved along in those fearsome currents. Since the Great Ice Age this stream of rock fragments has been grinding over the larger boulders and outcrops, carried along first in one direction and then the other. A few really massive boulders remain, and some of these may be erratics dumped by the melting glaciers. Slightly out of the mainstream, along the edge of the Narrows, the currents are less powerful. As a result this is often a complex zone of coarse boulders, rocks interspersed with outcrops of bedrock. In sheltered pockets there is even coarse sand. Paradoxically, because of the uneven nature of such areas, some very local turbulence effects can be even more powerful than those in the main flow. Life must be extremely rigorous for animals living on the tops of such boulders.

To the north of Ballyhenry Island, as the currents flow into the main body of the Lough and gradually slow down, the sediments change. Once deposited, large stones tend to remain there, giving rise to a rough, uneven sea bottom that eventually grades into areas of cobble-sized stones and coarse gravels.

The process continues. As the currents continue to slacken off, progressively finer deposits are laid down. Gravels are replaced by coarse sands. These, shifting back and forth in the ebb and flood currents, produce a series of lowlying 'dunes' that run at right angles to the current flow. Beyond, extending past Dunnyneill Island towards the Long Sheelah the sands become finer until a series of coarse muds, still mixed with rock fragments (and shell debris), dominates the scene. Finally, as we move away from the effect of the tidal currents (this means moving away from the central channel) the bottom becomes dominated by very fine muds and silts, where a single stroke of a diver's flipper can raise a cloud of particles that may take an hour to settle.

In the shallower areas (within the 10m depth zone) this pattern of sediment types is modified by wave energy. Anyone diving in any of the more exposed shallow areas can feel the wave surge push back and forth. The effect on the sediments is similar, and areas with some degree of wave action tend to have progressively coarser sediments closer to the shore zone. In the Narrows and between some islands, the combination of wave and tidal current energy described earlier can create periods of particularly violent wave action along those shores; however on most shores of the Lough the effect of the tidal currents is lost, and the predominant source of hydraulic energy is usually from wave action.

A rising tide in the north of the Lough, with the crag-and-tail of Scrabo Hill in the background. Materials eroded from glacial deposits form the main source of beach sediments. Even the most gentle wave action continuously sorts and redistributes these.

By their very nature, the headlands and rock outcrops often occur in some of the most exposed areas where finer materials have been carried away. At the entrance of the Lough, and along the eastern shore south of Kircubbin one can see where the different beds of Greywacke grits and shales, varying in strength, have been exposed and eroded at different rates and have thus formed a series of sharp ridges. In some places pockets of slight shelter have allowed the reverse to occur; coarse sand churning around has worn the strata down to a series of beautifully smooth curves - excellent examples of this can be seen at Mullog Point.

Elsewhere, the glacial boulder clays of the drumlins gradually crumble under the assault of waves and weather, constantly providing new material for the shores. This is the most widespread source of beach material in the lough. The south-west side of Dunnyneill Island is a classic case as it faces the main fetch of south-westerly gales driving up from the Quoile. At its foot lies a jumble of large boulders recently tumbled from the cliff and sheltered between them, an accumulation of finer material destined to eventually be carried off and deposited in more sheltered conditions elsewhere. Most of the other islands off the western shore of the Lough are not so exposed; they show the typical smooth drumlin shape that has suffered relatively little erosion since the last ice age. By contrast, what must once have been islands lying off the eastern shore have now been eroded into an intricate series of low shingle banks or

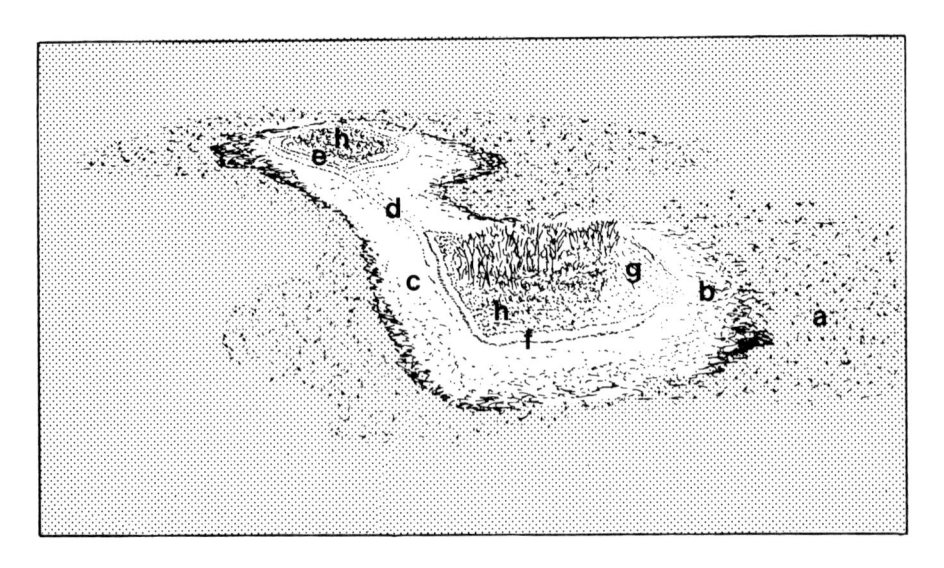

The anatomy of an island.

    a) Shallow subtidal zone of boulders overgrown by kelps;
    b) Low tide boulders and cobbles, subject to strong wave action;
    c) Sheltered shore of finer gravels and stones;
    d) Spit of fine shingle always being sorted, eroded, and redeposited;
    e) Acretion of very fine gravel, stabilized by strandline vegetation, and ideal for nesting terns and waders;
    f) One or more strandlines of weed and debris - rich in invertebrates;
    g) Rapidly eroding SW face, exposed to wave action from the Quoile area - probably supplying the island with much of its beach material;
    h) Raised beach areas overgrown by dense vegetation – typical nesting ground for wildfowl

pladdies, and their material of boulders, sands and finer sediments is now the property of the Lough's waters. Other eastern shores exposed to high levels of wave energy usually have cobbles or rounded shingles - many of the headlands between Greyabbey and Kircubbin, and again at Marlfield are like this.

    In the north of the Lough the largest areas of shore occur as vast expanses of muddy sand exposed for as much as 5Km out during low water. Although the area is shallow and relatively free from tidal currents, the considerable wind and wave fetch from the south means that conditions can still be quite choppy. The most obvious feature to look for is the pattern of ripples created by the wash and backwash of each wave running over the shore. These lie across the main direction of wave movement, but any small post or boulder in the sand will have a small pool scoured out, and a twist to the pattern created by the increased

Dunnyneill Island, near Killyleagh, viewed from the west.

turbulence. At the top of these expanses of shore the profile is usually steeper, and the upper fringes are usually composed of cobbles and shingles, collapsed from the glacial deposits when the finer materials are eroded around them.

Finally, in the most sheltered areas, particularly in the west of the Lough, we find those dangerously soft muds where even the birds appear to have trouble walking. Such areas are easily spotted by the smooth, glistening appearance of the mud, often indented by deep channels where streams flow across. If a stone is thrown, it sinks into the mud with a dull 'splat.' Clearly, these areas are not suitable for walking, but if you do get caught, try and get to a stream, where the sediments may be slightly coarser because of the increased water movement.

Out on the islands and pladdies, and on many of the headlands it is possible to spot extremely complex examples of the way the waves and currents interact with the land. At this point it is important to stress that not all the hydraulic energy from the Lough's waters results in erosion. Whilst the windward side of the land will probably be eroded, the sheltered side often trails off in a spit of fine shingle or coarse sand, revealing that the area is one where the waves or currents are losing some of their energy, and deposition of water-born materials is occurring. In some places (Dunnyneill Island, Gransha Point, or the Sheelahs) they may take the form of a long shingle ridge. These places are transitory features, and although they can justifiably be marked on maps, in reality they shift around, increase or diminish in complex responses to the movement of sediments in the latest storm. On occasions materials may be thrown up above the usual high water level, to be colonised by land plants and nesting birds; at other times, a storm with a new wave direction may sweep them away, back to the shore or into deeper levels. Vegetation on the shore itself may muffle the effects of wave action, and so reduce levels of hydraulic energy. In these conditions further deposition may take place, thus modifying the whole profile of the shore. In this manner the salt marshes are formed, as we shall see in a later chapter.

This is the restless setting for the Lough's rich wildlife. Above all, it is the way that the shape of the Lough as an area of sea has created a unique combination of forces operating through tidal currents, winds, waves and even small rivers and streams, each with its own materials and conditions. In doing so it has created an almost infinite variety of habitats in which wildlife can grow and flourish.

4 *Life Below the Waves*

IT SHOULD be clear by now that the waters of Strangford Lough can be dangerous, murky, and often stormy. Yet they harbour some of the richest marine life in north west Europe. Over two thousand different types of marine animal have already been found, representing about 70% of the species recorded on the whole Northern Ireland coast. 28 have not been found anywhere else on the Ulster coast. Moreover, since new discoveries are still being made we can be sure that the list is far from complete.

This array of animal life is not spread evenly throughout the Lough. We have already seen how much conditions vary from one area to another; equally, the different species within the Lough, with their varying preferences for certain conditions, show great differences in distribution. But the variety is on a much finer scale than that; when divers report that the animals living on a single boulder can vary even from one part of the boulder to another (for very good biological reasons) we can begin to realise that the marine life of the whole Lough is almost frightening in its complexity.

If variety is one important feature of the Lough's marine life, abundance is definitely another, for the bottom of the Lough and its waters are very crowded. In some areas it is impossible to see the bottom because of the tangled carpet of animal life. Brittle starfish may number over five hundred to the square metre, extending a forest of thin arms into the current to feed. Dense mussel beds may themselves be almost smothered by the animal life growing on their shells. Vast meadows of anemones, hydroids and sea squirts sway silently back and forth in the flow. Even the barren appearance of soft sands and muds is a deception, for close examination reveals a wealth of burrowing worms and shellfish living just below the surface.

These enormous populations are food for a variety of predators wandering over the bed of the Lough and in its waters. Starfish, crabs and even octopus take their toll, along with large numbers of fish which live on the bottom or swim in the water above. In turn, all of these may provide food for birds, seals, other marine mammals, and of course man.

In this way, feeding in the whole Lough operates as a unit, or a web of relationships. Changes in the fortunes of one species can have repercussions elsewhere in the system, but these are very difficult to predict, as our knowledge about the way the different species interact is still very limited.

Energy for Life

Virtually all life on land and sea obtains energy in the form of sunlight in order to live and grow. Plants harness this energy directly by using sunlight to convert carbon dioxide, water and basic nutrients into the carbohydrates, fats and proteins which are the main ingredients of all life. Animals cannot utilize sunlight in this way. They must obtain energy either by consuming plants, or eating animals which themselves have consumed plants. Thus plants play the key role in harnessing the sun's energy for all life in the Lough. The most abundant converters of the sun's energy are usually never seen. They are the countless microscopic plants that grow and flourish in every drop of seawater in the Lough. Along with the microscopic animals, they are known as the Plankton: plants are referred to as Phytoplankton and the animals, Zooplankton. Lacking virtually any means of active movement or propulsion, these tiny organisms are entirely at the mercy of the currents and waves. Swept in

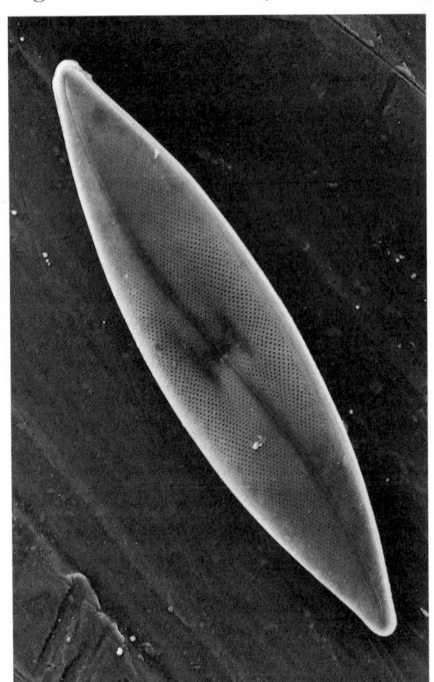

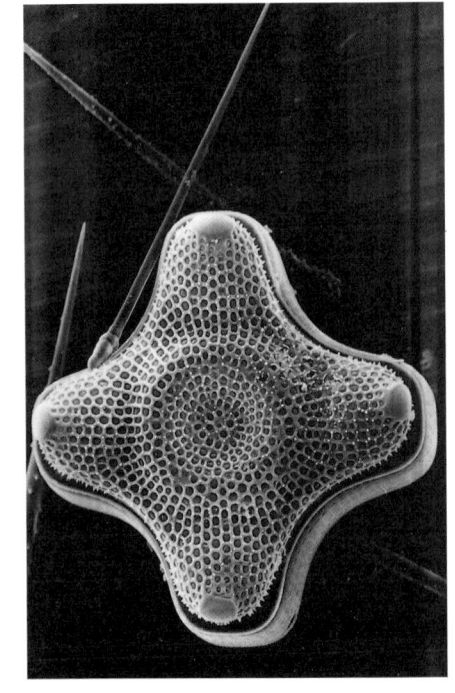

Diatoms from a sponge near Rathlin Island (x1100 and x550 respectively). These microscopic plants play a major role in the Lough's ecology (B.P.)

The Lion's Mane Jellyfish (Cyanea lamarkii), *one of the largest members of the plankton. Their tentacles frequently form a haven for small fish (B.P.)*

from the north Atlantic on the Gulf Stream, tossed around in the Irish Sea, they become caught in the tidal currents of the Narrows and carried into the Lough, which itself makes a major contribution. It has been calculated that about 700,000,000,000,000 of these microscopic plants are carried in and out of the Narrows with every turn of the tide; sufficient to turn the spectacular tidal currents from a mere physical feature into a vast swirling stream of life, reaching every inlet, depth and shore of the Lough.

Many of these tiny plants are extremely beautiful. Diatoms, less than a tenth of a millimetre in size, shaped like little pillboxes, have intricate sculpturing on their outer walls, creating beautiful symmetrical patterns. Others have a variety of elaborate projections, or subdivide to form complex chains of plants. Other types, known as Dinoflagellates, actually have a very limited swimming ability. They have no control over where they are carried, but they can orient themselves in the water and some can even attack and consume other microscopic organisms. Some of them are more characteristic of warmer waters and their appearance in our seas is usually during the warmer months of the year. A night-time visit in August or September can give a good idea of the abundance of this microscopic life. Find a dark spot in the south of the Lough, well away from street lights, and wade out into shallow water. With luck you may see, with every ripple or splash, a mass of tiny flashing lights in the water.

Alternatively, watch the turbulence from an outboard motor in the dark - the effect can be truly spectacular! This luminescence is caused by a warm water dinoflagellate called *Noctiluca scintillans* (or, translated more prosaically, 'flashing nightlight').

The quantity of this phytoplankton in the water varies considerably over the year. Stocks are lowest in winter when the water is at its coldest and daylight hours are short. With the onset of spring and increasing daylength and warmth, the diatoms, and shortly after the dinoflagellates, rapidly multiply and flourish, so that by May they give the water quite a cloudy appearance. In the open waters of the Irish Sea, this 'bloom' as it is called, is found only in the brightly lit surface layers and the rapid depletion of all the surface nutrients results in a decline in plankton levels in summer. In Strangford Lough the situation is very different. The turbulence in the Narrows redistributes the nutrients from the depths rather in the same manner that a brewer will help his fermenting beer by stirring it from time to time. As a result, productivity continues at a high level with a series of subsidiary blooms, and the Lough's waters therefore remain astonishingly rich throughout most of the year.

The animal component of the plankton, the zooplankton, is also very abundant, some 2,100,000,000,000 individuals passing through the Narrows with every flow of the tide. There are various species which feed on the phytoplankton, and equally there are a range of predators consuming other members of the zooplankton. Although the majority can reasonably be described as microscopic, being less than a millimetre in size, some types of zooplankton are enormous. These include the jellyfish, the commonest being the Moon Jelly (*Aurelia aurita*). Specimens of the Octopus Jelly (*Rhizostoma octopus*), well over two feet in diameter, quite frequently drift past the ferry slipway at Portaferry. Two other species, *Cyanea lamarckii* and *C. capillata*, known together as the Lion's Mane, are capable of delivering quite a nasty sting. In spite of their impressive size, jellyfish are scarcely more in control over where they are carried than are the very smallest members of the zooplankton, and they often end up on the shore where they are broken by waves or parched in the sun.

The most abundant types of planktonic animal are the crustaceans, the group that includes crabs, lobsters, and shrimps. The majority of planktonic crustaceans are copepods, tiny torpedo shaped animals. These occur in enormous numbers, and they are capable of consuming vast quantities of diatoms, particularly after the spring bloom. Although they are less than a millimetre in length, it is possible to see them in a bucketful of water scooped up from almost anywhere in the Lough. Look out for tiny specks swimming resolutely along in a rather jerky motion,

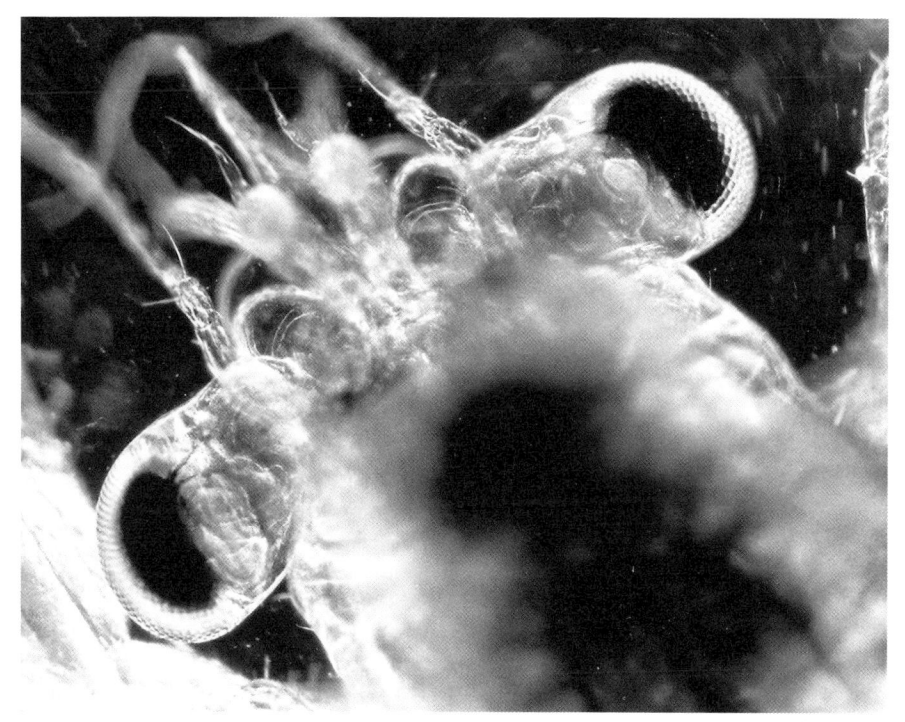

A crab larva (x200) from water near Portaferry.

pausing occasionally and sinking slowly. There are numerous different types, and collectively they outnumber almost all the other animals in the Lough, occurring in all its waters, on the shores, and amongst the weeds.

There are many predators in the zooplankton. Planktonic worms, small relatives of carnivorous species found on the shore, feed on the copepods. Occasionally it is possible to find the translucent Arrow Worms (*Sagitta elegans*) which are even capable of attacking fish larvae. Indeed, there are some carnivorous copepod species as well. However, the most frequently seen predators are the Sea Gooseberries (*Pleurobrachia pileus*), relatives of the jellyfish. They are one of the grand predators of this microscopic world, sometimes reaching a diameter of two centimetres. They are worth examining closely (scoop them up in a glass or jar), not least because of their beauty. Several rows of tiny iridescent hairs run the length of the animal and catch the light. These vibrate back and forth, and with the currents thus created, the animal can adjust its position in the water. Trailing behind are two long threads and these carry a battery of adhesive cells that trap smaller planktonic animals.

A very large number of animals in the Lough spend only part of their lives as plankton. In the same way a butterfly spends part of its life as a

Encrusting species growing at about 30m depth off Ballyhenry Point. Included are Dead-man's Fingers (Alcyonium digitatum), *a sponge* (Mycale) *and the anenome* (Actinothoe sphyrodeta) *(B.P.)*

caterpillar, so most of the animals found as adults on the bed of the Lough have lived as planktonic larvae before developing the adult form. These larvae are carried along in the currents, and so they are widely dispersed from the parent population. Any sample of water (particularly in summer, after many of the bottom dwelling species have spawned) is a soup containing the larvae of shrimps, crabs, worms, starfish, molluscs and even fish. In addition, the water may even hold the larvae of species not normally present as adults in the Lough (for example *Palinurus elephas*, the Crawfish) these having been carried in, possibly from as far away as the north Atlantic. For a variety of reasons conditions may not be suitable for their development so they never settle as adults, or having settled, they may be outcompeted by other species.

During the year the production of larvae by the populations of animals clustered on the Lough bottom is prodigious. One mussel alone can produce well over ten million eggs. The majority of larvae do not survive however, and a large proportion of those that do may be carried out of the Lough to be dispersed further afield. It is a debatable point as to whether the Lough receives more larvae carried in from the Irish Sea, or exports more larvae produced by its own animal populations. Two things seem clear however. Firstly the Lough's capacity to provide habitats for an extraordinary variety of animal types, in great abundance, means that there is a rich storehouse of plankton-producing wildlife that can act like

Life Below the Waves 59

a reservoir for the less productive conditions outside in the Irish Sea. Secondly, the availability of enormous quantities of plankton as food is the main reason why the Lough can support such exceptionally dense populations of bottom dwelling animals. Implied in this last point is the fact that the Lough system allows an overproduction of larvae so that many species both produce plankton, and obtain it from the water as food.

Life on the Bottom

Most of the different types of animal that live in the Lough are found on the bottom, from depths of sixty metres through to shallows stirred by wave action at low tide. It is a scene that has continued to develop in the Lough for some 12,000 years since the end of the Great Ice Age. Post-glacial changes in sea level, coupled with increases in the water temperature, allowed the Lough to develop as a rich marine system. However most of the plant and animal types are far older than this, many probably dating back to late Tertiary times. As the post-glacial changes in the Lough progressed, these would have successively colonized the new areas of sea bottom available.

What is this ancient world like? Although it is almost silent, it is full of movement. A constant stream of plankton and tiny particles drifts over the bottom. Below depths of about ten metres, there is little light and few

Brittle Stars living on boulders in Marlfield Bay. Good water exchange supports dense populations of these filter feeders (B.P.).

plants can grow. Only the most persistant wavelengths of light, the greens and blues, can filter through. All the other colours are lost in the dim twilight of half shapes in the cloudy water. A bleak prospect, perhaps, but all the diver needs to do is switch a torch on! In the small pool of torchlight, unimpaired by passing through metres of water, the true colours of the Lough bottom suddenly explode, revealing a crowded, animal-dominated world of extravagant reds, pinks, whites, yellows and iridescent violets.

Temperatures vary between about 5°C in winter to 16°C in late summer. This is a much smaller range than we experience on land (about -1°C to 20°C) and the change from one day to the next is almost imperceptible. With the exception of places where streams enter the Lough, the salinity of the water remains at a roughly constant thirty-three parts of salt (and other minerals) for every thousand of water. Thus for animals living on the bed of the Lough, conditions may be regarded as generally mild and relatively predictable. It is important to remember however, that animals adjusted to such a regime may be much less tolerant of changes than their counterparts on land. Nonetheless, the relatively equable conditions have allowed populations on the bed of the Lough to develop and exploit the rich feeding conditions to a much greater extent than would be the case if they suffered the vagaries and climatic extremes that beset terrestrial life.

With the almost endless supply of drifting plankton and other nutrients, it is not surprising that an enormous proportion of the Lough's bottom dwelling animals feed on this material, and do so by filtering it. These filter feeders provide one of the most important means by which energy gained by plankton, swirled down from the sunlit waters above, can be trapped and exploited in those poorly lit depths where the sun's light is minimal and plants cannot survive. Subsequently, the filter feeders themselves fall prey to other predators and so this energy is passed on through the system.

Animals on the bed of the Lough have a remarkable range of devices for filter feeding. Some possess tentacles. The Plumose anenome (*Metridium senile*) for example, has a dense and beautiful crown of feathery tentacles extended into the current. The colonies of the soft coral Dead-man's Fingers (*Alcyonium digitatum*) use a similar method: every individual, or polyp, has its own crown of tentacles, each covered with stinging cells that trap planktonic organisms. As each finger-like projection may contain several hundred of these polyps, an area of bottom covered in soft corals represents a formidable means of trapping drifting food. Many of the brittle stars operate on a similar principle, this time by extending their arms into the current; these trap the plankton

Life Below the Waves

and, with the aid of a stream of mucus running down each arm, sweep the food into their mouths. Fan worms have a crown of tentacles on their heads, which is similarly used to trap plankton.

Other animals use more active methods of filtration. Often this involves use of the gills, which play a dual role of respiration and as a mesh for trapping plankton. Bivalve molluscs open the two halves of their shells slightly to allow the gills to pump water in and out of the shell cavity. Sea squirts, bag-like animals with two siphons, set up a filtering current. Action by microscopic hairs draws water in through one siphon, passes it through a sieve-like pharynx, which traps the food, and pumps it out through the other. Some animals operate on the principle of the fan, or rake. Barnacles for example, collect plankton using highly modified feathery legs, and it is possible to see the feeding process in action by watching barnacles in a shallow pool, their tiny bristles sweeping through the water, drawing food into the shell cavity.

The filter feeders are only one aspect of the complex web of life on the bottom. Predatory species like crabs, the Curled Octopus (*Eledone cirrhosa*), fish and seals have a major role to play as well. Equally, the scavengers, species like the Whelk *Buccinum undatum,* and a number of the shrimps and crabs help to break down the remains of dead animals and plants. Many of these act as both carnivore and scavenger, depending on what food is available. Deposit feeders consume small fragments of organic detritus that settle on the bottom. Bacteria occur everywhere, particularly in the sediments, and complete the breakdown of organic matter.

*A Spiny Starfish (*Marthasterias glacialis*) on Lee's Wreck, Ballyhenry (J.B.)*

Grazing species like the sea urchins feed by scraping off microscopic organisms growing on stones, boulders and bed rock.

The various roles played by the different species equally apply to many other areas of shallow sea. It is the sheer variety of physical conditions which has created such an enormous range of habitats, and which makes Strangford Lough outstanding. Each of these habitats supports a particular selection of species that can live together in the local conditions, and among them will be the filterers, the predators, the grazers and scavengers. All these combine to form what may be loosely described as a community. If we move around the Lough, and thus into different habitats, it is likely the communities will be different, each holding a different selection of species performing these various roles, although the more tolerant types may feature in a wide range of habitats and communities.

Animal Communities

To see the way the different animal communities are spread over the bed of the Lough, it is best to start in the Narrows. A visit to the bottom in this area will reveal outcrops of bedrock and massive boulders virtually carpeted by brilliantly coloured sponges and soft corals. The few species

A male Cuckoo Wrasse (Labrus mixtus) *off St John's Point, Donegal. The species is often seen in the Ballyhenry Wreck (B.P.).*

Life Below the Waves 63

capable of surviving the powerful currents exploit the rich feeding to the full. The grey Elephants's Ear Sponge (*Pachymatisma johnstonia*) can grow to the size of a large beach ball, and the giant form of the brilliant yellow Boring sponge (*Clione cellata*) may be almost as big. Everywhere, the projections of the soft coral Dead Man's Fingers, ranging in colour from orange to white, support clusters of tiny polyps; approach too closely, and they all withdraw back into the body of the coral.

Amongst the boulders, particularly away from the fiercest currents, there may be a bigger range of species. Some of these very uneven areas may be extremely turbulent, but there is some compensation in the form of thousands of clefts and small holes sheltering more delicate species. The hydroid *Tubularia indivisa* is particularly common in these areas, often forming a dense forest of small tubes and tentacles extending into the current, frequently being eaten by sea slugs.

At this stage, the wreck in Ballyhenry Bay near Portaferry ('Lee's Wreck') deserves special mention. Viewed from a boat or from land, she is merely an ugly hulk of rusting iron, but originally she was one of the hastily constructed 'Liberty ships' built in the U.S.A. to replace shipping losses in World War II. Called the Empire Tana, her life in service was brief and after the war she was towed into the bay with the aim of scrapping her on the shore of Ballyhenry Island. Fortunately for us all she got stuck on a rock and as the tide fell she broke in two. The rear section still lies where she went aground, and although she is now in a dangerous condition, and collapses a little bit more each winter, the old Empire Tana has become a paradise for marine life.

As you dive down the dim walls of the wreck, obscured by a constant blizzard of small particles and plankton, the ability of marine life to colonize new sites rapidly becomes obvious. The old iron hull is literally smothered in thousands of Plumose Anenomes, ranging in colour from white to a rich traffic-light amber. Many are over four inches across, crowned with a mass of feathery tentacles stretched out to trap the food streaming past. Elsewhere, colonies of Dead-man's Fingers mingle with sponges and sea squirts. Groups of the transparent Light-bulb Sea Squirt (*Clavelina lepadiformis*) swing back and forth in the current. Common Sea Urchins (*Echinus esculentus*) graze on algae and detritus on the bottom, and in the nooks and crannies, the aggressive Velvet Swimming Crab (*Liocarcinus puber*) lurks. Take hold of its shell crosswise with finger and thumb (carefully !) to see its paddle-shaped rear legs thrash back and forth as it tries to swim away.

The wreck is full of fish. In summer tiny Sand Eels (*Ammodytes tobianus*) flicker in and out of the structure in their hundreds. Small bottom dwelling fish like the Father Lashers (*Myoxycephalus scorpius*) swim over

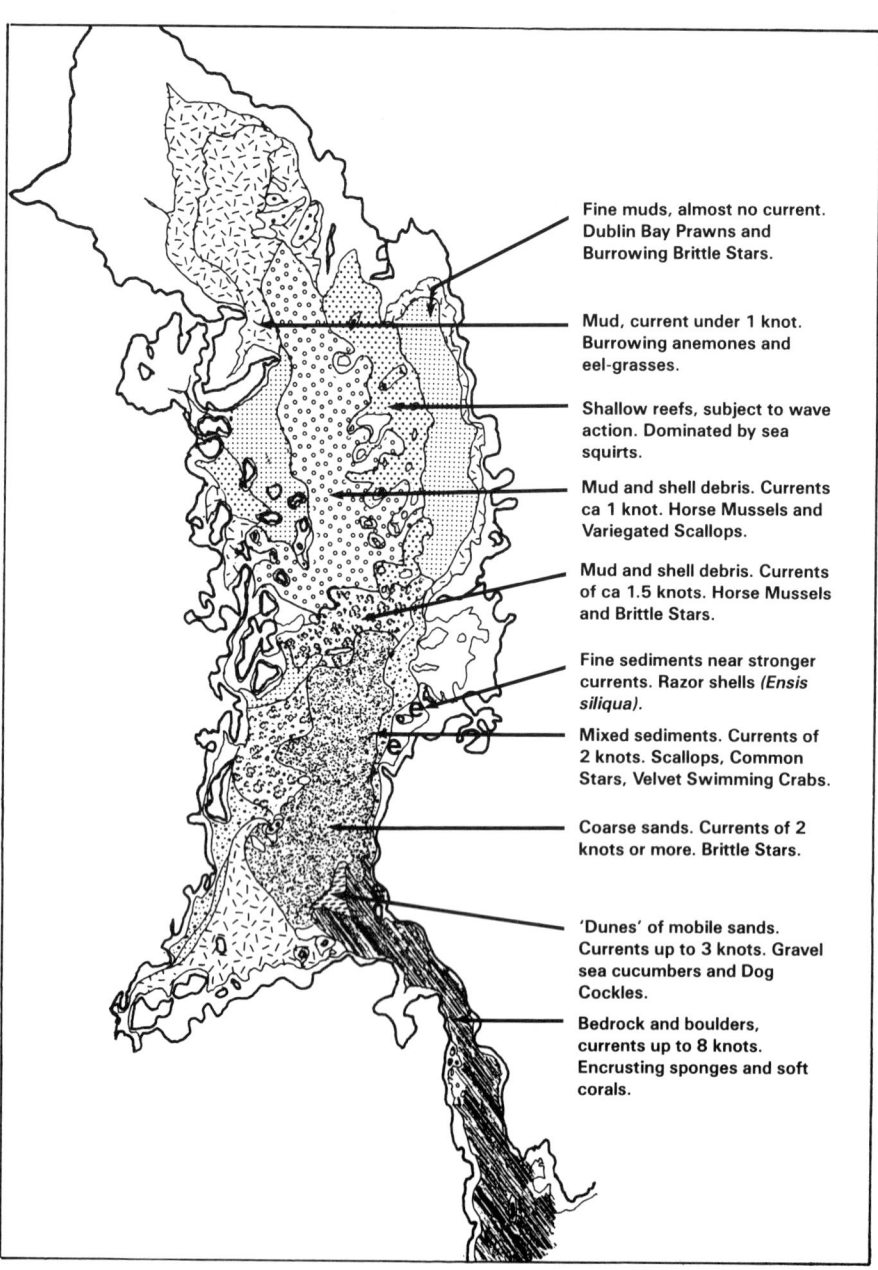

The main subtidal animal communities in Strangford Lough.

the bottom or scuttle along the hull. Both Cuckoo and Ballan Wrasse (*Labrus mixtus* and *L. bergylta*) can be seen, often coming to investigate the diver. The former start life as reddish females in a group, with one dominant male. If this male disappears for some reason, then the most dominant female changes sex, increases in size, and takes on the characteristic male colours of the most beautiful irridescent blue. Many of these fish and the occasional Conger Eel (*Conger conger*) can actually be seen within the structure of the wreck, but divers should take particular care, for it is still collapsing.

To the north of Ballyhenry Island, with a gradual reduction in currents and changes in sediments, a succession of different comunities can be seen. Boulders grade into broad areas of cobbles, intersperced with patches of very coarse sands and gravels. The most characteristic animal is the anemone *Actinothoe sphyrodeta* - white tentacles and yellow central disc - but it is also possible to see the pink crusty growths of calcareous algae in the more brightly lit shallow areas. Strangely enough, these deep high energy areas are also home for the Butterfish (*Pholis gunnellus*), which is more commonly regarded as a sheltered shore species.

In some of the deepest parts, to the North of Ballyhenry Island, the 'dunes' of very coarse mobile sand are one of the first habitats in the succession where burrowing species become really important. It takes an observant eye in these very dim parts of the lough, but it is possible to make out the pale branched tentacles of the Gravel sea cucumber (*Neopentadactyla mixta*) - though on occasions the whole population has the annoying habit of simultaneously withdrawing below the surface for no obvious reason, leaving the area apparently devoid of life! It is in this area that the beautiful shells of the Dog Cockle (*Glycymeris glycymeris*) may be found, just barely poking through the sand.

Progressing further into the main body of the Lough, roughly east of a line from Jackdaw Island, Dunnyneill and the Limestone Rock, up to the Long Sheelah, the power of the currents begins to decline. Deep areas of rather flat, open sands are almost completely obscured by vast quantities of brittlestars. As one descends to the bottom, the millions of thin arms extended into the current in the dim greenish light look almost like a field of grass. By far the most numerous species is the Common Brittlestar, (*Ophiothrix fragilis*) but anywhere in the Neill reef area, or in other areas where the current may be slightly less, look out for the Black Serpent-star (*Ophiocomina nigra*) amongst them.

A rich variety of other species lives in this area. Among the more obvious ones, we can find the Great Scallop, or Clam (*Pecten maximus*) once abundant, but now grossly overfished. The smaller Queen Scallops (*Aequipecten opercularis*) may also be present. Predators like the Curled

Octopus, and various species of starfish regularly occur. The spectacular Sunstars (*Crossaster papposus*), in various shades of red, and often as big as soup plates, are probably the most numerous, along with the ubiquitous Common Stars (*Asterias rubens*). The fate of their prey is an unenviable one. The starfish are equipped with lines of tiny tube feet on the underside of each arm. Each foot is equipped with its own tiny sucker, and with them the starfish not only moves over the bottom, but grasps prey such as scallops. The starfish enfolds the victim in its arms and slowly pulls the shells apart until the gape is wide enough for it to extrude its stomach into the shell cavity, and literally digest the victim alive. Unless the prey is strong enough to resist the attack, the only defence is one of speedy escape before the attack begins. Find a Queen Scallop and a starfish on the bottom and gently touch the former with the starfish. The scallop will violently open and close its shell like a set of false teeth, the resulting jet of water propelling it up and away from danger.

In the western and central areas of the Lough as far north as Greyabbey, where the currents are much reduced, one of the most interesting and certainly the richest habitats is provided by dense populations of its main species, the Horse Mussel (*Modiolus modiolus*). This is the largest species of mussel in the British Isles, rounded in shape, coloured dark brown or black and unlike its counterpart on the shore, inedible. The extraordinary life style of this rather drab beast could fill a book, and there is an

A Curled Octopus near Ballyhenry Point; a notable predator in a number of subtidal animal communities (B.P.).

The Horse Mussel community with Variegated Scallops (Chlamys varia) *near the Long Sheelah. One of the richest communities in the Lough, it is being severely damaged by bottom trawling (see inset) (B.P.)*

increasing body of opinion that regards the Strangford Lough Horse Mussel beds as being unique.

Three features of the Horse Mussel make it special. The first is its ability to live securely in very soft mud by attaching itself to the buried dead shells of generations of former mussels and other shellfish. Initially the mussel community must have had some firmer basis to become established, since mud without such shell debris is unsuitable. The community must therefore have established itself when there was less fine sediment, and more stones on which to become attached. This would have required greater water movement, and such conditions could have occurred during the periods of higher sea level some 7,000 years ago when there were probably additional outlets to the Irish Sea via Dundrum and Kirkistown on the Ards Peninsula. Today's rich Horse Mussel community, in areas of the Lough that are now unusually sheltered for this species, could well be a relic of these ancient conditions.

The second feature is their long survival. Small, young mussels die by the thousand from attacks by crabs and starfish. But the few that manage to achieve a size where they are immune to these attacks, may live for over forty years. These old mature individuals form the bulk of the population, and since they have grown big enough to survive attacks, they have diverted most of their energy away from growth and into long term reproduction.

The abilities to live in soft mud and to live for a long time combine to create the third special feature; an extremely rich habitat. Mud is almost impossible for encrusting species like sponges or barnacles to grow on, but the hard shells of these venerable mussels protruding above the surface of the mud provide them with stable surfaces for attachment. Likewise, hard rock is out of the question for most burrowing species, but the channels of soft mud between the clumps of mussels is ideal. In this way, this single community of mussels can allow a far wider range of species to become established than almost any other, and over a hundred different species have been found associated with the Horse Mussels.

At first glance, the mussel community is a bewildering confusion of animals. The encrusting and surface species catch the eye first, every mussel clustered with hydroids, sponges, barnacles, sea squirts, and the tangled calcareous tubes of polychaete worms. Variegated Scallops (*Chlamys varia*) are particularly numerous, and this feature has not, to my knowledge, been found in any other Horse Mussel community in the British Isles, but it does occur to a limited extent in Scandinavia. The older the mussel, the more overgrown it is, until only the slight gape of shell drawing in plankton-rich water reveals it as a living animal. In the silt between the mussel clumps and tangled in the shell debris we find the Smooth Porcelain Crabs (*Porcellana longicornis*), Scale Worms (*Lepidonotus squamatus*), sea cucumbers, and even the colourfully striped Football-jersey Worm (*Tubulanus annulatus*) that may be as much as a metre long. If some of the mud is examined in the dark, it may give off flashes of brilliant green light - caused by a tiny brown worm, impressively called *Flabelligerina affinis*. Many of these burrowing species are feeders on dead or decaying matter, deposited on the surface of the mud, between the particles, and ejected as faeces by the mussels. Indeed, the great majority of the animals are microscopic, living on this food between the individual particles of mud.

The mussel community has a number of mobile predators and scavengers. Spiny Squat Lobsters (*Galathea strigosa*) lodge between the mussels for saftey. Anglerfish (*Lophius piscatorius*), Curled Octopus, and starfish are quite numerous, and there are several crab species. In addition to the large Edible Crabs (*Cancer pagurus*), Common Hermit Crabs (*Pagurus bernhardus*), and the Strangford Hermit Crab (*Pagurus cuanensis*), there are large specimens of the Great Spider Crab (*Hyas areneus*), and of the Scorpion (*Inarchus dorsettensis*) and Long-legged (*Macropodia rostrata*) spider crabs. Both camouflage themselves by attaching small pieces of weed or sponge to their shells - in the dim light of twenty metres down this disguise is most convincing. The sponges may

Life Below the Waves

also give the crabs extra defence by being extremely unpalatable to predators in their own right.

Where the Horse Mussel beds rise towards the central boulder pladdies of the lough, the cluster of relic islands and shallow reefs provides even more shelter from the main current flow. In these areas, the mussels gradually give way to dense colonies of sea squirts, (*Ascidiella aspersa*) gently wafting back and forth in response to waves some way above at the surface.

Finally, there are those depths of the lough off the Western shore, or sheltered to the east of the central pladdies, where water movement is virtually negligible and where the very finest silts come to rest. Gone are the rocks and boulders, or mussels that provide a base for the encrusting animals of the lough. Instead the bottom appears flat and almost lifeless. Look more closely, however, and it is possible to see that the whole surface is pockmarked with tiny holes and tracks. Some of these may be the signs of the animal known variously as Dublin Bay Prawn, Norway Lobster or Scampi (*Nephrops norvegicus*). This handsome (and tasty) animal typifies the burrowing lifestyle of these areas, creating complex systems of tubes running through the mud, with numerous entrances. Perhaps the most characteristic view of this beast is of it guarding the entrance to its burrow, claws extended menacingly at the intruder.

An Angler fish, photographed in the Northern Ireland Aquarium at Portaferry

A Dublin Bay Prawn on its habitat of soft mud. This species is typical of areas with the weakest tidal currents (B.P.).

There are many other species dwelling just below the surface. A burrowing brittlestar *Amphiura filiformis* can just be detected as a faint circle of tentacles as it feeds on microrganisms in the sediment. In some areas numbers of the burrowing anenome *Cerianthus lloydii* extend tentacles from their tubes into the water. The Pelicans-foot Shell (*Aporrhais pespelicani*) is another sediment dweller (also seen in the Horse Mussel community). Turn it over to see the bright pink snail within. In some areas it is possible to find Slender Sea Pens (*Virgularia mirabilis*), looking like a series of old fashioned quill pens stuck in the mud. Related to sea anenomes and soft corals, each 'pen' is a colony of polyps with one long individual forming the stalk and the remainder living on the branches. We find a division of labour here; some polyps are equipped with tentacles for trapping planktonic animals, whilst the rest play the role of breathing, pumping water and supplying the whole colony with dissolved oxygen. These beautiful, but very fragile animals are highly luminescent - if stimulated, phases of light pass upwards through the length of the colony.

If there is one key point to be drawn from this summary of the animal life on the bed of the lough, it must be the sheer abundance of the whole system. Not a single spot remains unoccupied; when a new site becomes available for occupation (as with the sinking of the Empire Tana at Ballyhenry) it is immediately colonized by animal life settling out from the soup of planktonic larvae in the water. These enormous populations of animals require a massive daily food intake, and their success is ample evidence that they get it. Were it not for the rich nutrient supply, the turbulence of the Narrows circulating the plankton carrying their store of the sun's energy to the depths, and the ability of the filter feeders to trap and harness this energy, the Lough would be a much poorer place.

Fish of the Lough

Such a range of animal communities might lead one to suspect that the fish populations would be equally spectacular. This is indeed the case, but we know much less about them than the other aspects of the Lough's wildlife. This seems strange in the light of the Lough's commercial fishing importance, particularly during the last century.

Some fish are integral members of the different bottom communities. They can be seen everywhere in the Lough, scuttling over the bottom, up boulders, and amongst the holdfasts of kelps in the shallower areas. Being small, vulnerable, and not very mobile, they rely considerably on camouflage. The dumpy little Father Lashers for example, combine a nobbly distorted form with a mottled brown colouration which makes them almost indistinguishable from the sea bottom, unless they move. Their squat shape may also help them to hold their ground against the powerful currents. In the north of the Lough, particularly in the shallow areas of muddy sands, Plaice (*Pleuronectes platessa*), Dabs (*Limander limanda*) and Flounder (*Platichthys flesus*) feed on items such as the exposed ends of worms burrowed in the mud. In the rich kelp growths, immediately below the island shores, enormous numbers of Eels (*Anguilla anguilla*) hunt for small crabs, worms and shrimps. When the tide is in, they migrate up into the shore areas to feed amongst the boulders and weeds.

One bottom dwelling species to look out for particularly is the Angler Fish. Often over a foot long, it has an enormous mouth armed with an array of dangerously sharp backwards-pointing teeth. Its dirty brown colouration blends with debris on the bottom so well that it is almost invisible. The one conspicuous feature is a long thin dorsal fin ray, swollen at the tip to look like a morsel of food. Any small fish approaching this bait is allowed to investigate for a few seconds, and then that enormous mouth opens up and snaps shut over the prey.

The mid-water depths of the lough contain the large shoals of the more mobile pelagic fish, kept buoyant by gas filled swim bladders within the body cavity. They are also equally at home near the surface, or picking small prey off the bottom. Among the former are Herring fry (*Clupea harengus*) and Sand Eels which feed on the Lough's rich plankton stocks.

These finger-length silvery fish arrive in their millions throughout the warmer months of the year, and if they are chased by Mackerel (*Scomber scombrus*) from below, the water seems to boil as they rise and erupt at the surface catching the light. The best time to see this is in summer or early autumn during a rising tide, in the areas of turbulence some way to the south east of Dunnyneill Island. Not only is it possible to see the Sand Eels exploiting plankton-rich upwellings and the Mackerel in pursuit, but it is worth keeping an eye out for seals, terns and Porpoise fishing in the area as well.

Cod (*Gadus morhua*) start life by feeding on shrimps and small crabs, and as they grow they graduate to more active prey like Herring and Sand Eels. Herring used to be plentiful in the Lough (a local name for them is 'Portaferry chicken'), and Lough-side agricultural communities relied on a good catch to supplement their income. Occasionally there was a glut; one farmer apparently even attempted to fertilize a field near Audley's Point with surplus Herring, but according to local wisdom he only succeeded in poisoning the land for several years, presumably because of the excessive oil. Those days are now long gone, and the Herring stocks, possibly because of overfishing, have not been regarded as worthy of regular commercial exploitation for some fifty years. Coalfish (*Pollachius virens*) are numerous, but their main claim to fame lies in the fact that at different stages of their lifes they have different local names, being variously called Saithe, Gilpen, Blochen, Glashan, Greylord, and Coley.

One species that appears to have maintained its numbers reasonably well is the Grey Mullet (*Chelon labrosus*). A visit to many of the small rivers entering the Lough during high tide, when the salt water penetrates upstream, may reveal some of these fish swimming in to feed on algae and small animals. The stream at Castle Espie is a classic example of this. Sea Trout (*Salmo trutta*) also inhabit various areas, and the best way of seeing these is to go to a sheltered shore during low water and watch for a splash and a swirl in the quiet waters as they hunt for small fish.

Much larger fish species also occur, although lately their numbers have declined considerably. The Lough was once famous for its massive Skate (mostly *Raja batis*) well over 2m in length and weighing about 90Kg, but along with all the Irish Sea populations these have almost disappeared. Nonetheless, Tope (*Galeorhinus galeus*), small sharks of up to 1.5 m in length, still occur from time to time, the females coming into the Lough to 'pup.' At the entrance of the Lough, particularly to the east of Gun's Island, and sometimes in the Narrows, it is possible to see the second largest fish in the world, the truly massive Basking Sharks (*Cetorhinus maximus*). In spite of their awesome size, (sometimes up to 13m long)

A Basking Shark off Peel, Isle of Man. They are often seen in summer at the entrance of the Lough, and occasionally in the Narrows.

they are totally harmless to man, feeding on rich surface stocks of plankton, which they filter at rates of well over 1,480 cubic metres of water in an hour. They are best seen during slack water between the tides, in spells of warm summer weather, each with its triangular dorsal fin breaking the surface, followed by the tail fin swinging slowly from side to side.

The Ultimate Predators

Strangford Lough is well endowed with predators taking fish and the larger bottom dwelling animals, and they can be regarded in some ways as the final stages of the whole network of feeding relationships in the Lough. Energy derived from the sun has been harnessed by the phytoplankton, which have been consumed by the zooplankton and bottom dwelling filter feeders; these in turn have been taken by bottom predators and fish. Now it is the turn of the birds, seals, Porpoises and even the occasional whale. Because they are dependent on all the other components, these animals are often regarded as good indicators of the 'health' of the system.

A number of birds typical of the open sea come into the Lough to exploit its food resources and to enjoy comparatively sheltered conditions. They are important members of the Lough's marine life, living almost exclusively on fish. In summer Sandwich (*Sterna sandvicensis*),

Common (*S. hirundo*), and Arctic (*S. paradisaea*) Terns are the most ubiquitous of the sea birds; virtually every corner of the Lough has some of these graceful white birds diving recklessly into the water for Sand Eels. We shall be looking at their nesting islands later, but the important point at this stage is that although they can fly as much as twenty four kilometres to obtain food for their young, those nesting in Strangford Lough have a rich supply of food on their doorstep, making the Lough an ideal nesting area.

Cormorants (*Phalacrocorax carbo*) and Shags (*P. aristotelis*) both occur on the Lough in large numbers, though the latter are most usually seen in winter. These two species, together with Red-breasted Mergansers (*Mergus serrator*), exploit shallow waters, frequently round islands, and over shore areas covered during high tide. Here they catch the wrasse, Eels, and other fish which move in amongst the dense bands of shallow kelps. Often it is possible to see a Cormorant struggle with an Eel thrashing about, but held securely in its bill. Red-breasted Mergansers are particularly suited to catching slippery fish, since their bills are sharply serrated. In some inshore waters, particularly around Whiterock, Slavonian Grebes (*Podiceps auritus*) occur, hunting small fish and crustaceans.

Out on the open waters there are large numbers of auks, a group of

Common Seals in Granagh Bay, south of Portaferry. The Lough's population is the largest of any site in Ireland.

species that appear to have particular trouble taking off from the water, which would seem to be a major disadvantage for a sea bird. They also dive for fish, and it is then that their true abilities become apparent. Their poor performance in take-off is in fact a compromise, for the birds are remarkably adept at 'flying' underwater, the specialized shape and structure of their wings allowing them to pursue fish with incredible agility. Black Guillemots (*Cepphus grylle*) are particularly numerous in late summer, whilst Razorbills (*Alca torda*) and Common Guillemots (*Uria aalge*) occur throughout the year, sometimes in large flocks moving around the Lough, presumably in response to fishing conditions.

Occasionally, especially after winter storms, sea birds from much further afield visit the Lough. When severe weather affects western Scotland, a trip to the Narrows can be rewarding. Large numbers of Gannets (*Sula bassana* - sometimes over 700 birds) and Kittiwakes (*Rissa tridactyla*) can be seen diving for fish in the waters round the Angus Rock, or merely cruising into the wind low over the water. Other species may appear from time to time; Manx Shearwaters (*Puffinus puffinus*) regularly patrol the entrance to the Narrows, gliding along the troughs between the massive waves of the Bar. Red-throated Divers (*Gavia stellata*) sometimes frequent the waters off Killard Point. Storm Petrels (*Hydrobates pelagicus*) and Great Northern Divers (*G. immer*) can turn up further into the lough, and there are even stray Puffins (*Fratercula arctica*).

The most familiar carnivores in the Lough's waters are the Common and Grey Seals (*Phoca vitulina*, *Halichoerus grypus*). From the Angus Rock at the entrance, to almost as far as Newtownards, they bask on the rocks exposed during low tide. Individuals can move rapidly from one area to another, but they tend to be quite conservative in their choice of haulout sites, so that once a spot is established as a haulout it tends to remain so for some time, holding anything up to a hundred seals. Shortly after high tide, as the ebb gathers pace, they can be seen taking positions on rocks covered by a few inches of water. Gradually the water level drops, exposing them to the air. Often there appears to be a certain amount of jostling for the best positions, but eventually they settle down and rest, unless they are disturbed, until the next rising tide. There are many places to watch them; for those travelling by car, the Cloghy Rocks National Nature Reserve presents the best opportunity, but they may also be seen in Granagh Bay N.N.R., Barr Hall Bay, and various other sites in the Narrows. For boating people, Green Island Rock, and the various pladdies off Dunsey Island near Ringhaddy are ideal, although caution is required by those with large boats, and unneccessary disturbance should be avoided.

Common Seals are by far the most numerous, in 1987 numbering

A newly born Common Seal pup at Bankmore, in July (J.B.)

about eight hundred; probably the largest accumulation in Ireland. This probably reflects the enclosed nature of the Lough as they are regarded as being very much a sheltered coast species. Numbers reach a peak in June and July when many give birth to their pups. These can often be seen on the haulouts at this time, coloured a metallic blue-grey with a dark streak down their backs (Common Seals shed their white coats BEFORE birth). Within a few hours of birth the pups are quite capable of swimming with their parents, their smaller heads seen bobbing within a few feet of the mother. At high tide, when the haulouts are covered, the majority disperse to hunt for food. Although they have quite good eyesight, it is thought they hunt fish by detecting vibrations in the water, probably using their long whiskers as antennae. Often they come in close to the shore, partly to inspect any human visitors, for seals are notoriously inquisitive, but mainly because there is good fishing to be had. Those fresh water streams with Grey Mullet moving to and from the deeper water may well have a group of seals nosing around for the chance of making a catch. In the Autumn, young seals come in to play; a boisterous affair involving great leaps out of the water, loud splashes snorts and growls. The best spot for watching this is in the bay immediately to the south of Cloghy Rocks, during high tide, but there are many other places in the Lough where it occurs. Courtship and copulation is much more discreet and takes place underwater.

Grey Seals are more characteristic of open exposed coasts and there are fewer than fifty in the Lough at present, although recent years have seen an increase. They can often be seen alongside Common Seals, but their greatest concentrations are on rocks in the Narrows, and the central pladdies near Bird Island. These are usually the areas where pupping occurs, although one actually gave birth on the shore opposite the 'Scotsman' in Portaferry. The very fluffy white pups (they moult this fur some time AFTER birth) are usually born between September and November, and because they are often reluctant to swim for the first fortnight or so, they give the misleading impression of being approachable and even deserted by the mother, who in fact may only be a short way off in the water. Beware! A seal pup approached by an inquisitive human is a very frightened animal armed with an excellent set of teeth.

Both species of seal are amazingly versatile in their selection of food. Their numbers are viewed with concern by fishermen who resent their consumption of Salmon (*Salmo salar*), Sea Trout and other commercial species. Whilst they undoubtedly do take these, the seals also take a much wider range of non-commercial animals like bottom dwelling fish, crabs, and shellfish. Indeed it has also been suggested that they may actually have a beneficial effect by taking Eels, which consume the young of some commercial species, and Octopus which prey on young Lobsters. The real truth is that the picture remains unclear as to what the effect of the seal population actually is. However, two things seem clear. Firstly, the seals will only remain whilst there is adequate food for them, and if they deplete the food stocks excessively they will have to move elsewhere. Secondly, the seals are a major tourist attraction, and increasing numbers of visitors are coming to watch them.

A Grey Seal pup on the Craiglee Rocks south of Kircubbin. Although they seem deserted at this time, the mother is usually nearby.

In the Autumn of 1988 the Lough's Common Seal population suffered from the disease, now called Phocine Plague, that had already killed up to twelve thousand seals on the coasts of north west Europe. The speed with which the disease spread was astonishing, with the first carcasses being found within about a fortnight of the outbreak in England. By the end of the outbreak over 100 victims of the disease had been found, but the annual survey conducted the following year revealed a reduction in numbers of about 200, or about one third of the 1988 population, although not all of this may be due to the plague. The effects of the disease on the seals were particularly malignant, impairing breathing, swimming and hunting ability, and finally causing paralysis, ensuring a painful death drifting in the water and subject to attacks by gulls. Careful watches for any reoccurence of the plague are maintained by conservation organizations, but so far this has not happened, and the Lough's population currently fares well, producing about a hundred pups each summer.

The majority of the large mammals seen on any trip are likely to be seals, but a number of other species may occur, particularly in the south of the Lough, where areas of turbulence attract the fish. Porpoises (*Phocoena phocoena*) often appear in schools of up to ten, but it is difficult to be certain of their numbers as they do not all appear at the surface at the same time. Their movement through the water is very different from

Life Below the Waves

that of seals, and once seen, there is little possibility of confusion. They travel rapidly, the triangular dorsal fins clearly visible as they break the surface. Usually they are wary of boats, but occasionally an individual will come close. Such moments are truly memorable as they dive and rise just in front of the bow, often seeming to look up out of the water at the curious humans peering down on them. The vent on the top of the head gives a snort as the animal exhales before the next dive, and the horizontal tail flukes driving the porpoise through the water are all clearly visible from the boat. These are the smallest of the British whales, only reaching a length of about two metres. More occasionally, Bottle-nosed Dolphins (*Tursiops truncatus*) visit the Lough, and there has sometimes been some confusion over which species has been seen, although they are considerably larger than the Porpoise.

Sometimes the Lough is visited by larger whales. Pilot Whales (*Globicephala melaena* - also called Blackfish, Caa'ing Whales, and locally, Herring-hogs) occur quite frequently off Angus Rock and round to Gun's Island in summer. Often five to six individuals circle shoals of Herring into tightly packed concentrations of confused fish before moving in for the kill; fishermen use this behaviour as a guide to the best fishing spots. Once every few years Killer Whales (*Orcinus orca*) enter the

A Porpoise surfacing near Dunnyneill Island. Occasionally they will accompany boats for some distance.

Lough, the last documented occasion being in July 1982 when seven came in and remained for about a day. They were a spectacular sight, the largest being about ten metres long, the massive dorsal fins of the males protruding high above the water, the stark patches of white on their flanks showing clearly. There have been no recent records of any other species of whale; in the past however, larger species may have ventured in, as their bones, presumably collected as souvenirs in the last century, still lurk in some older gardens round the Lough. Perhaps the best known was a pair of whale's ribs forming a decorative arch in the garden of a cottage in Marlfield, north of Portaferry, until they finally rotted away in the 1970's.

Towards the Shore

As we move away from the depths and the open waters of the Lough into the shallows just below the shore, the pattern of tidal currents remains similar, although in some places they may be an even more powerful. Many of the animal communities are also similar, for many of the species

---

CONSERVATION OF THE SEA BED AND WATERS OF THE LOUGH

Unlike all the other features of the Lough, this most fundamental aspect of the Lough at present has virtually no protection or management. Most of the bed of the Lough is owned by the Crown Estate Commissioners, who have the authority to control exploitation, through extraction of sea bed materials or moorings, for example. However, planning regulations only reach down to the low water level, not below, and therefore use of the seabed and waters of the Lough does not come into the Local Area Plans. The bed of the Lough is in effect a planing void, often subject at random to what happens on the surrounding land and shores.

Fisheries managment is under the control of the Department of Agriculture Fisheries division. They are aware of the unique importance of the Lough's subtidal marine life, but have had no policies to cater for this, in spite of the fact that work is in progress to designate the Lough as a Marine Nature Reserve. As a result considerable damage has been done to the marine life on the bottom of the Lough, to the detriment of both fishery interests and the conservation of the marine life.

Although designation of the Lough as an M.N.R. would be welcome, the limitations of the legislation mean that an overall plan for the whole Lough and its hinterland, supported by the community, is essential if it is to be effective.

from deeper water can thrive quite successfully in the shallows. There are however, some important differences. In these shallow areas sunlight penetrates to the bottom, and therefore plants can grow, in many areas developing into luxuriant forests of kelp and red algae. At the same time these are the levels most subject to turbulence from wave action on the surface, so that the sediments are constantly washed back and forth. Even in areas well removed from tidal currents, most of the finer particles are removed if there is any degree of wave action.

Thus as we approach the shores all round the Lough, and round its islands and pladdies we find a brightly lit, complex, restless, but narrow zone of shifting sediments, swinging thickets of kelp, and often a bewildering mixture of animals, where representatives from the shore above co-exist with other species more at home in the deep waters below. In areas of boulders and bedrock, the wash and backwash can be quite violent, and streams of bubbles sweep through the weeds, as air is trapped by the breaking waves. In more sheltered waters, gentle growths of thick bladed eelgrass sway in shifting sands. Fish swirl around in the water foraging for small animals amongst the weeds, whilst diving duck, sea birds, seals, and even Otters (*Lutra lutra*) hunt for food. Above lies the shore, where the marine life comes to terms with an additional, and very different range of conditions.

5   On the Shore

SEASHORES ARE often dismissed as thin strips of ground lying between land and sea, regularly flooded by the latter. Technically, this is correct, but the real picture is much more exciting! Ireland's shores form a negligible part of its total land area, yet they are endowed with an enormous proportion of its habitats and wildlife, as well as magnificent scenery. Strangford Lough goes one stage further; fully 40% of the Lough's area is intertidal, in other words, lying between the high tide and low tide water levels. Moreover, with some 240 Km of shore, the Lough holds an astonishing 37% of Northern Ireland's 650Km coastline.

The shore life, just as in the subtidal areas, is exceptionally rich. The variety is due mainly to the contorted shape of the Lough, and the varying materials that make up the shore, whilst its abundance is largely due to

Butterfish near Roe Island; the species is common on many shores, as well as in some of the deepest parts of the Lough (B.P.).

the same features that contribute to the wealth of life below its waves. This is scarcely surprising, since virtually all shore animals and plants are truly marine species, and depend on the sea for their survival. In particular, the rich nutrient and plankton content of the lough's waters supports the dense growths of shore plants and animals along the shore. As a result, the filter feeding animals that feature so much in the subtidal parts of the Lough also have their representatives in all the different habitats of the shore.

But there is another reason why the Lough's shores, and indeed most shores, have such a great variety of life. In many habitats (meadowlands for example) the areas that combine the greatest fertility with the most 'ideal' conditions do not neccessarily have the greatest variety, although some species may be very abundant. It is often in the habitats presenting the greatest challenges, where over millions of years, animals and plants have struggled for survival, that the greatest diversity of wildlife occurs. On Strangford's shores, the fertility of the Lough's waters is countered by challenges posed by a number of very different habitats. These habitats are all united by the constant rise and fall of the tides, subjecting the plants and animals to a cycle of environmental changes that in any other wildlife context would be regarded as catastrophic.

Another big difference between the Lough's shore and its depths is that the shore receives enough light each day to sustain plant growth. Apart from the shallows immediately below the low water mark, the depths are almost entirely devoid of plant life, but the rock or boulder shores have luxuriant growths of seaweed smothering the rocks with a dozen shades of brown. The mudflats in the north of the Lough, although seeming rather barren from a distance, also have a range of their own algal species, whilst in summer and early autumn they are tinged with the green of the eel-grasses. All these plants play a fundamental role in shoreline communities.

Problems for Shore Life

Nearly twelve and a half hours separate one high tide from the next. During this time almost every part of the shore will be exposed to wind, rain, sun, heat or cold. The plants and animals experience extremes of environmental conditions that are far in excess of those affecting their relatives in the depths of the Lough. This period out of water thus represents a considerable challenge to their survival, and they have all devised ways of either overcoming the problems or circumventing them in some way.

The most obvious risk to any shore species is that of drying out when

Dense weed growths (here on Ballyhenry Island) protect more delicate animals from desiccation during low tide.

uncovered during low tide. Plants resist desiccation by producing abundant quantities of mucus, or slime, which helps to keep them supple and moist, even in the sun. This characteristic makes them notoriously slippery for us to walk on, but it is extremely important to the hoards of worms, crabs, anemomes, sponges and sea squirts which shelter in the damp conditions. Animals living on sandy and muddy shores with little weed growth, burrow into the sediments to stay moist. In fact, a very large proportion of the animals are structurally incapable of resisting dessication, but they survive exposure of their habitat by virtue of their behaviour. The most widely used strategies are either to retreat from such conditions whilst the tide is out, or in the case of sedentary species, to select suitably damp locations in which to settle and grow.

Other plants and animals resist the hazards of drying out with physical features like thick shells or physiological adaptations. Limpets (*Patella vulgata*) and barnacles, with their robust conical shells lose very little water when the tide is out. Some snails can seal the entrance to their shells with a horny plate, the operculum, that effectively blocks off any water evaporation. At the top of most rocky shores around the Lough, immediately below the strandline of weed tossed up by storms, there may be a line of a small brown seaweed which often looks dead, being very crisp and dry. This is the Channelled Wrack (*Pelvetia canaliculata*) which is capable of drying out almost completely (during neap tides for several days at a time) and then very rapidly reabsorbing water, thus 'coming to life' when the tide covers it once more.

In addition to the risks of dessication, there are large temperature fluctuations on the shore. As we have seen, the Lough's waters experience a gentle variation from about 5-16°C over the year. Air and land temperatures may vary more than this in a single day, whilst the yearly regime may have a range of about 21°C. Thus the retreating tide may suddenly expose the shore to the extremes of a very hot summer's day, or to the frost of a clear winter's night. Then later, an abrupt return to sea temperatures with the rising tide. In very cold spells, the shores can even aquire a layer of ice built up by freezing surface waters left behind during a succession of low tides. Animals of the shore (except mammals and birds) cannot control their body temperatures, so they avoid these extremes by burrowing or hiding under stones and weeds. Worms for example, retreat deeper into the sediments during very cold weather. Some species can tolerate rather more variation in temperature than others, and predictably the species dwelling at the top of the shore, experiencing the longest periods of exposure to heat or cold, are the most tolerant.

Salinity varies in a similar way. Upper shore pools can become very salty, as summer warmth evaporates them, whilst long periods of rain may render them brackish, or even almost fresh. Streams and rivers like the

Sheltered muds in Ballymorran Bay near Killinchy. Streams on the shore carry a range of wildlife adapted to constantly changing salinity.

On the Shore

A Heron (Ardea cinerea) *hunting on the shore at Gull Rock, Mahee Island (L.J.T.).*

Comber River, the Blackstaff, and the Blackwater, may immerse the local shore life in completely fresh water during low tide. When the tide comes in again, conditions suddenly become saline once more, and these abrupt changes can impose considerable stress on the body chemistry of many of the animals and plants.

Shore life in some parts of the Lough must also be able to withstand the high levels of wave action. Often these conditions are most intense on rocky or boulder shores, and whilst some soft bodied animals may thrive in clefts and holes, many animals receive the full brunt of waves hitting the rocks. Weeds like the kelps growing on rocks have to be pliable - almost leathery - to swing with the wave surge, rather than to resist it. In places these dense growths may help to damp down the effects of wave action. In areas where wave action is too strong for the plants, conditions are too extreme for more delicate animal types, and such areas are dominated mainly by animals with tough shells.

For a very small group of animals the tidal cycle of alternate exposure and inundation carries the reverse risk. Some marine animals living near the top of the shore have become so adapted to long periods out of water that they may actually drown if submerged too long. The most easily seen of these is the Sea Slater (*Ligia oceanica*) crawling over high shore stones and almost every pier and jetty round the lough. This little animal is well adapted to breathing air through moist gills, although in all other respects it is dependant on the sea for its existence. In addition, a very small number of insects have managed to colonise the shore, although they are typically a terrestrial group. One of these is the Bristle-tail (*Petrobius maritimus*), a close relative of the silverfish that one finds in old houses. Another is *Lipura maritima,* a tiny pale blue insect that can be found floating in rafts of up to fifty individuals on the surface film of water in sheltered, high-shore rock pools or under stones and boulders. Both species are really land animals and are dependent on air for survival. They retreat into crevices and sediments when the tide comes in.

As if all these different factors were insufficient test of the abilities of animals and plants to adapt to severe conditions, the Lough's shores are also exploited by a wide variety of predators. During high tide, fish from deeper waters and from the shore itself take worms, crustaceans and shellfish. The fish populations themselves may also be subject to attack by Cormorants, Shags, Red-breasted Mergansers, Dabchick (*Tachybaptus ruficollis*), Kingfishers (*Alcedo atthis*), seals and even Otters. Many of the shore's own predators - crabs, shorefish, and anenomes are also in action during this time. Then, as the tide recedes, a new wave of predators move on to the exposed shore. Shorebirds, particularly in winter, arrive in Strangford Lough in very large numbers, as we shall see later. They consume enormous quantities of small animals, green algae, and eelgrasses to survive the winter. Foxes (*Vulpes vulpes*), Badgers (*Meles meles*), and Brown Rats (*Rattus norvegicus*) all make use of the shore, particularly when cold spells freeze areas inland. On other occasions, Red Deer (*Cervus elephas*), Pheasants (*Phasianus colchicus*), and Pigmy Shrews (*Sorex minutus*) have all been found feeding on the shore. Man himself is a predator, particularly taking Lugworms (*Arenicola marina*) for bait, and shellfish like Cockles (*Cerastoderma edule*), Shore Mussels (*Mytilus edulis*), and snails.

Shore Communities

Virtually every surface of rock, boulder, stone, and the finer sediments on the shore is occupied by some form of life. Even those areas apparently bare of life generally support important communities of microscopic animals, plants, and bacteria. Because the great majority of shore animals are marine in origin, the range of filter feeding, predatory, and scavenging lifestyles is essentially similar to those found in most of the subtidal areas of the Lough. Plankton makes a major contribution to the energy of the system, and so we still find many filter feeders on the shore. It is not the only source of energy however, for the dense growths of large weeds, eel grasses and microscopic algae also harness the sun's energy, and support large populations of grazing animals whilst alive, and when dead break down into fine detritus, to be consumed by deposit feeding species. Other animals in their millions rely on coarser debris, for example the rotting mounds of weed cast up by storms. In addition, the shores are often the first parts to receive materials washed off the land. Much of this is a perfectly natural consequence of rainfall and erosion, but it is also true that the shore is particularly prone to the effects of sewerage, farm chemicals, and other effluents.

As in the subtidal situation, each type of animal or plant has its own preferred range of conditions, and therefore habitats. Thus the wildlife of the Lough shores, like its counterpart below the waves, is distributed in various assemblages, or communities, of different species. Each of these communities is very much determined by the physical nature of the shore, and just as in the subtidal situation, the degree of water movement is the primary factor. On the shore this is largely derived from wave action, and Strangford Lough's complex topography, combined with those strong and frequent south westerly winds, means that the wide range of levels of exposure to wave action results in a correspondingly wide selection of different communities.

There is however, a vertical dimension to the communities on the shore, that greatly increases their complexity. Shoreline species vary in

Serrated Wrack (Fucus serratus), *a seaweed typical of lower levels of the shore.*

A rocky cleft in Silurian bedrock at Killard Point. Such features provide pockets of shelter in an otherwise exposed rocky headland.

their abilities to withstand the extremes of intertidal existence, so their distribution in relation to tide level varies considerably. Those species most capable of withstanding exposure are aligned at the top of the shore. Those least capable (but often very competitive in other respects) appear at the bottom end of the shore, and between, a selection of species thriving at various levels. These are arrayed in much the same way that the subtidal species ordered themselves according to the strength of the tidal currents. Thus between high and low water are a series of communities, each composed of the range of species most successful at the various levels. Because of the way in which the large seaweeds are arrayed in relation to tidal levels, these conspicuous plants appear as well defined zones, clearly seen on many shores.

The materials of which the shores are built play a major role in shaping the communities to be found about the Lough. Admittedly the materials themselves depend to some extent on the other factors; one would not expect soft muds to be found at the extremities of an exposed headland, for example. Equally, freshly eroded glacial material is more likely to occur at the top of a beach rather than further down, when it has already been sifted by wave action. But within any set of conditions of exposure and tidal level the different materials present exert their influence. The orientation of folds and beds in the rock, the presence of igneous dykes, the size and shape of the boulders, the variations in sediments deposited by the glaciers, all affect the selection of animals and plants, or the community that lives in each locality.

Exposed Rocky Shores

No areas in the Lough ever receive the full force of a really good Irish Sea storm to the extent experienced by places like the islands off Portavogie, or the Maidens off Larne. Nonetheless, there is sufficient wave action to produce the conditions which result in exposed shore communities. These are based on the Silurian greywacke grits and shales. Because these outcrops are most in evidence poking through the thinner glacial deposits in the south of the Lough, they are under the influence of the tidal currents as well. Most exposed are the outer rocks at Killard and Ballyquintin Points, and on the Angus Rock, which are truly spectacular places to visit in a storm.

Such sites have a somewhat restricted plant and animal life. At the top of the shore (and on many shores about the Lough) the black lichen *Verrucaria maura* is the first true marine plant to be encountered, appear-

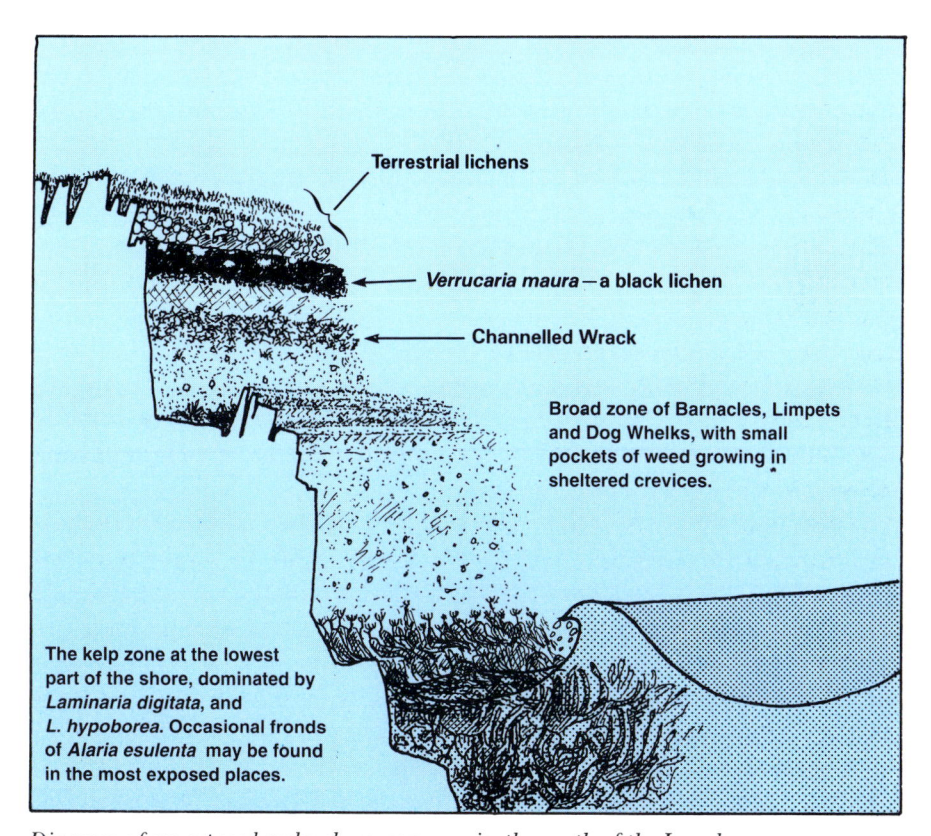

Diagram of an exposed rocky shore, common in the south of the Lough.

ing as a thin black film growing on highshore rocks that are only covered briefly at high tide, or may even only be subject to wave splash. In these exposed places, the spray may allow it to grow a short way inland. It often occurs close to the terrestrial lichens *Calloplaca* and *Xanthoria*. These brilliant yellow-orange species provide a delightful contrast to the black of the *Verrucaria* a few inches further down. There may be a few other top shore plants in such situations, most notably *Pelvetia canaliculata*, and another lichen, the spongy looking *Lichina*.

Below this level lies a zone usually bare of plant life. Few animals live on the open stretches of rock except dense populations of barnacles (two species here- *Balanus balanoides* and *Chthamalus stellatus*), and Limpets. During high tide the Limpets laboriously scrape the surface algae off the rocks with their teeth-bearing abrasive tongues, before finally returning to the 'homescars', a closely fitting depression each has ground into the rock in which it lodges during low water. Because their shells fit tightly against the rock face, the limpets are able to avoid dessication. Their grazing plays a significant role in preventing plant growth becoming established, and in turn this deprives more delicate species of the shelter they need during low tide.

At the low water zone, the kelps are found, broad bladed weeds that may grow to three metres in length, with large holdfasts; root-like structures that adhere to the rock surface. In the most exposed rocks at the entrance of the lough, the leathery blades of Dabberlocks (*Alaria esculenta*) with their distinctive yellow midrib, may be found. Everywhere thick forests of Kelps (or Oarweeds) *Laminaria digitata*, and slightly below, *L. hypoborea*, provide habitat and security for large populations of small fish and shrimps amongst the fronds swinging back and forth as the long open-sea swell washes round the rocks.

In some of the sheltered clefts produced by the folded rock beds and the igneous dykes, a colourful range of other species like the anenome, *Sagartia*, as well as hydroids and sea squirts like *Dendrodoa grossularia* can occur. The Dog Whelk (*Nucella lapillus*) is common in such areas, and it is a major predator of barnacles and Limpets, drilling small holes in their shells and rasping away the flesh within. It is occasionally possible to find the Edible Crab in the larger overhangs and gullies near the low water levels. Both green and red algae can be found, but the most notable are two distinctive species of red algae. They differ from all the others in that their tissues are heavily impregnated with lime. *Corallina officinalis* is delicate pink with finely divided, very brittle branches, growing in a fringe about two inches thick round many of the rock pools located in bedrock, well sheltered under the weeds and stones. The other is *Lithothamnion*, growing as a hard purple-red crust over rocks and stones,

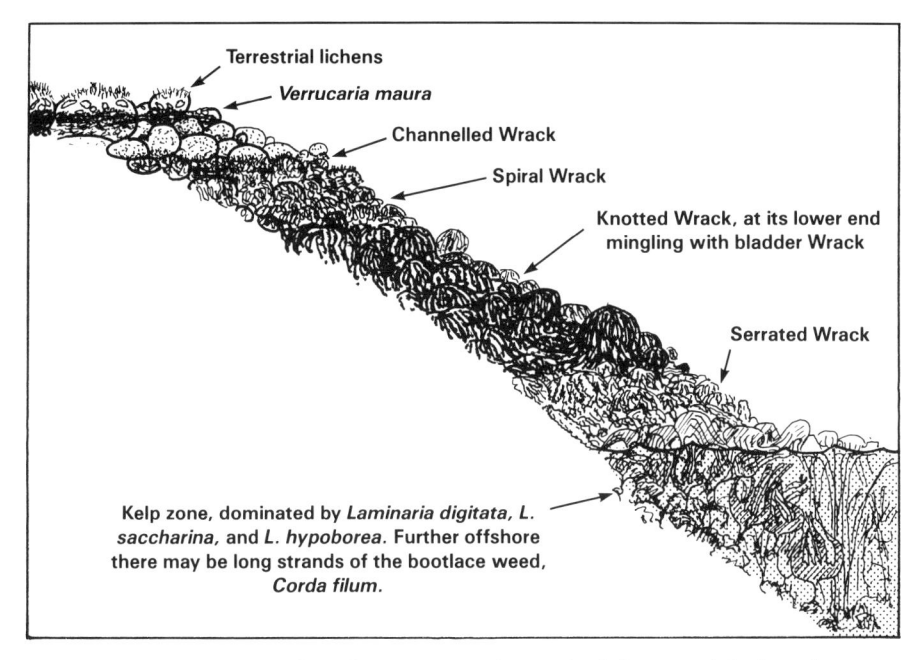

Diagram of a sheltered boulder shore, common in much of the west of the Lough.

sometimes completely smothering the latter. Most exposed areas have this species, both on the shore and in the shallow subtidal areas. Often it is possible to find specimens of both these rather strange plants washed further up the shore, but sadly, within a short period of their death, both species lose their beautiful colours, fading to a dull grey-white.

Apart from species found sheltered by these corners and hollows, the most exposed rocks do not carry as rich or abundant a wildlife as more sheltered localities. The nature of the exposed shore community is perhaps more dominated by the catastrophic effects of the few really severe storms (and the species that can survive these) than by the more usual moderate conditions prevailing during most of the year.

Sheltered Rocky Shores

Further into the Lough, the situation changes rapidly. Places along the length of the Narrows like Audleys Point, Ballyhenry Point, the rocks at Kilclief, the Gowland Rocks off Granagh Bay, the outer Cloghy Rocks, and Swan Island opposite the Strangford Ferry terminal, all experience protection from severe wave action. Moreover, these outcrops and headlands protect other areas of rock; even slight differences in aspect and situation provide the conditions neccessary to support very sheltered shore communities, in spite of the powerful currents. Within the main body of the Lough, similar areas occur along the eastern shore, for example at Black Neb and Rowreagh Point to the south of Kircubbin, although the plants and animals still have to withstand waves driven across from the Quoile area. Along the western side, on the islands off Ringhaddy and Whiterock, and as far north as Mahee Island, as well as on many of the pladdies in the centre of the Lough, steep boulder shores derived from drumlins continue the theme. Protected from the prevailing winds by the high drumlins on the adjacent land, these are often even more sheltered, and supply countless clefts and crannies between the boulders.

All these hard, but sheltered shores of rock and boulders are rich in marine life. Stable surfaces for attachment, and sheltered conditions with a good water supply, allow dense growths of brown algae to develop, beautifully illustrating the zoning of the various species into different levels. Look along any of these shores during low tide, and it is possible to spot the zones as clear lines of various shades of colour and texture. These thickets of brown algae are among the richest examples of this range of habitats to be found in the British Isles.

At the top of the shore, we have already encountered the Channelled Wrack with its extraordinary ability to survive dessication. This level tends to have a rather restricted range of the few animal species that can survive

Laminarians growing at the lowest level of the shore off Kircubbin. These are some of the most exciting areas to explore for shore life.

The arctic starfish Leptasterias mulleri, *extremely rare in Northern Ireland outside Strangford Lough.*

the long periods out of water, but often they occur in large numbers. There are often lots of small dark snails, the Rough Periwinkle (*Littorina saxatilis*). This animal is well adapted to living long periods out of water, having a type of lung modified from an enlarged shell cavity lining; moreover, it stores its eggs in this moist interior, until they hatch as fully formed young, each protected from dessication by its tiny shell. Other species cope in different ways. Amphipods hide under stones and pebbles, as well as occurring in vast numbers in the moist piles of rotting weeds along the strand line.

The next plant species to be encountered is the dark-brown Spiral Wrack (*Fucus spiralis*), followed by extensive areas of the Knotted Wrack (*Ascophyllum nodosum*) with its air filled bladders. These keep the weed buoyant at high tide, and in very sheltered areas it forms a floating, oar-tangling fringe, at its lower end often mingling with the Bladder Wrack (*Fucus vesiculosis*). At low tide the weeds lie draped over the rocks and boulders in dense, shining, and slippery banks. Both species are extremely variable in appearance, some varieties being hardly recognisable. A curious and extremely rare variety of *Ascophyllum*, for example, can be found at Ballyhenry, Darragh and Rainey islands. Called 'variant *Mackii*'; it lacks the typical buoyancy bladders (so it stays sunk on the bottom during high tide), lacks the usual holdfast attachment to the shore, and grows into small bushy clumps that lie loose on sheltered areas of mud and stones.

The most frequently seen snail is the Smooth Periwinkle (*Littorina obtusata*) occurring in a range of brilliant reds, yellows, oranges, greens and browns. Less equipped to survive out of water than its top shore relative, it tends to avoid light, and this habit naturally takes it into the cover provided by the dense weed growth, as it grazes the surface algae off the stones. Amongst the crevices and weeds there may also be numbers of the Edible Periwinkle (*Littorina littorea*), the largest of the periwinkles in the Lough, and locally referred to as 'Whelks' or 'Wilks'.

This is the highest level where we usually find the shore fish living in the shelter of the weeds. Many species like the Shanny (*Blennius pholis*), various types of goby, and the Butterfish (known locally as Peter Nine-eyes), are well adapted to staying moist during low tide by producing large quantities of mucus. This habit also has the useful by-product of making them extremely difficult to catch (just try getting hold of a Butterfish!). Some species like the gobies and the Two-spotted Sucker (*Lepadogaster bimaculatus*), living in areas with some wave action have highly modified pelvic fins that act like a sucker, enabling them to grip the rocks.

Other species to look out for include the Dog Whelk, Beadlet Anenomes (*Actinia equina*) and the Common Shore Crab (*Carcinus maenas*). This latter species deserves mention at this stage, because it is, by a wide margin, the most widespread of all the larger shore animals. It thrives in an enormous range of conditions and depths, from the most sheltered corners of the Lough through to areas buffeted by wave action. It feeds on living animals, plant debris, and scavenges on dead or dying animals as well. When children run off the shore, with jam jars full of exciting discoveries, these tough little animals are almost invariably in the haul, and usually the ones best equipped to survive the subsequent handling and prodding!

Continuing down the shore, next lies a zone of varying width, of weeds coloured a somewhat lighter brown, the fronds having a characteristic saw-tooth edge to them. This is the Serrated Wrack (*Fucus serratus*), most frequent in the more sheltered interior of the lough. Finally, towards the bottom of the shore, close to the low water level, lies a complex zone of kelps. Within the lough, on places like Dunnyneill Is., the Limestone Rk., and many pladdies to the north, *L. hypoborea* is often dominant, thick growths glistening just above the water during low tide. Other kelps include the mushroom shaped growths of Thongweed (*Himanthalia elongata*) which develop into 2-3m thongs during the summer, and the broad blades of Furbelows (*Sacchoriza polyschides*) whose bulbous, warty-looking holdfasts are cast up by storms in winter.

Although the microscopic phytoplankton are, by a wide margin, the most important converters of the sun's energy into a form usable by animal life in the Lough, the brown algae also contribute significantly to the system. They remain largely uneaten whilst alive, but their broken fragments, tossed up by storms to accumulate and rot on the strandline, or deposited as fine particles of detritus, support vast populations of small animals. As a result the dense growths of weed on the western shore of the Lough, by virtue of the prevailing winds, must contribute significantly to the animal populations on the eastern shore where most of the coarser debris accumulates, and to those of sheltered mudflats where the finest particles settle.

The red algae are widely present on the Lough's rock and boulder shores, usually growing amongst the larger brown algae. They are some of the most beautiful of the shore plants, occurring in a spectrum of different reds and purples, and in many shapes and forms. They grow on rocks, on the larger brown algae, and in rock pools and sheltered clefts in rock faces. A few of them, most notably Laver (*Porphyra umbilicalis*) and Dulse (*Palmeria palmata*) are edible, the latter quite tasty in a salty kind of way if left to dry on a boat deck for a couple of hours. During the summer small boats, usually from Portaferry, can be seen making their way down to the sheltered, but rocky parts of the Narrows. Assisted in their passage by the ebb tide currents, they arrive in areas like Cloghy Rocks, Angus Rock, and the Pladdy Lug to hook the delicacy out from amongst the kelp growths laid bare at low water, before being carried back to Portaferry by the incoming tide.

A kelp holfast on Ballyhenry Island. Numerous holes and passages in the structure provide an important habitat for many low shore species.

The enormous variety of red algae precludes description of all of them, but some are particularly notable. *Delessaria sanguinea* and *Rhodymenia* both occur low down on rocky shores, usually away from the most severe wave action, amongst the rockpools and pladdies exposed during low water. Over the more exposed shores it may also be possible to find Carrageen, the short tussocky growths of *Chondrus crispus* and *Gigartina stellata*, the former with its lovely violet iridescence. (This disappears if the plant is removed from the water) These occur throughout the Narrows, on Ballyhenry Island for example, and on many more sheltered boulder and rock shores in the Lough.

At the lowest levels of the shore, the variety of animals is at its greatest, and so these areas are especially exciting places to explore, particularly during low water spring tides. On bare patches of rock there are often numbers of a small animal, with a segmented shell, adhering tightly to the rock. These are chitons, rather primitive molluscs, that can grip irregular surfaces and mould their shell to the shape of the rock for protection. There are large numbers of a small starfish, *Leptasterias mulleri*, coloured green by algae living in its skin. Although this species has been recorded on other parts of the Down coast, it is particularly characteristic of Strangford Lough. Its widespread appearance in the Lough is interesting, because it is more typically a sub-arctic species, more at home in northern Scandinavia and Iceland than here, to the extent that it is not even referred to in most books on British shore life. Instead of spawning and leaving its young to the dubious chances of a planktonic existence, *Leptasterias*, like many other northern species, broods and protects its young. Thus throughout the winter months the females hide under clefts and boulders, with their numerous tiny offspring held securely under their arms until they are able to fend for themselves.

Many animals use the dense weed growth as habitat in its own right, as well as shelter from dessication. A number of species for example, grow on the fronds, benefiting from access to a good water supply as the weed swings in the water. Bladder and Serrated Wracks are spotted with the tiny white spiral shells of tube worms, that emerge to trap plankton when the weed is covered during high tide. Small Hydroids like the pink *Clava squamata* grow on frond intersections on the Knotted Wrack. On the kelps there are usually small patches of lacelike growths, the sea mats (Bryozoans), colonies of tiny animals living in a complex network of calcareous capsules. It is worth looking out for the small cavities in the kelp stems made by the Blue-rayed Limpet (*Patina pelucida*), a particularly attractive mollusc with iridescent blue lines running along the shell.

The holdfasts, those root-like projections that attach the plant to the rock or boulder, provide a habitat for another selection of species,

The sea urchin Psammechinus miliaris, *common on sheltered rocky shores.*

sheltered from wave action and dessication by the weed growths above. The Smooth Porcelain Crab, a lively little purple-brown animal, is a frequent resident of these cavities. Scale worms, sea cucumbers, small molluscs, and sponges can all be found lurking within kelp holdfasts, creating a sort of micro-community within the shore habitat. Where there is good water movement, sponges are particularly numerous. The Breadcrumb Sponge (*Halichondria panicea*) is widespread, growing on holdfasts, weed stems and on stones. It is said to smell of freshly baked bread when broken, but to me it seems more reminiscent of garlic! The smell is quite distinctive, however, which is helpful in identification, since it occurs in a variety of shapes and colours. Another sponge quite frequently seen is *Myxilla incrustans*, easily distinguished from the former species by its bright orange-yellow colour and its tendancy to grow on the rocks rather than the weed itself.

Another habitat is provided by the sheltered clefts and crannies in the layers of grit and shale in the bedrock, and by the almost infinite number of spaces amongst the boulders. Some of the holdfast dwellers may occur here, but other species can also be found. The cushion star *Asterina gibbosa* is a common species under most boulders, varying in size from a couple of millimetres to about three centimetres across. Squat lobsters, *Galathea squamifera*, are very common; when disturbed they are capable of swimming rapidly, their tail flicking back and forth, propelling the animals backwards. In these areas another species of Porcelain crab occurs, the Broad-clawed (*Porcellana platycheles*) which is beautifully camouflaged to look like a small piece of rock, being flat and able to fold its claws and legs close into its body. It is also well adapted to holding on to the rocks; lift one up and offer your finger for it to cling onto - its tenacity is surprising for such a small animal. Sea squirts are common in this habitat, particularly *Dendrodoa grossularia*, appearing as a small reddish nodule on the underside of rocks, its two siphons quite clearly visible. Interesting fish can be found amongst these lower growths; several species of Blenny, rather roundheaded and brightly coloured, can be found. Another species to look out for is the Rockling (*Onus mustelus*), with five sensory barbels on the front of its head, that probably help in finding prey.

For many people the most exciting animal to be found is the Otter, and although it is described here, it could equally have been discussed in other parts of the book, for Otters also play a role in several subtidal, intertidal, and land habitats. This beautiful and versatile animal occurs throughout most of the Lough, and although there is at present little information to go on, I get the impression that the population is quite large by contrast with many other areas. Indeed, in the All Ireland Otter Survey of 1980/81, all but one of the sixteen potential sites on the Strangford Lough and Ards coast had signs of the presence of Otters. However, the majority of reports have come from areas of sheltered rock and boulder shores, particularly in enclosed inlets.

Although they are not often seen, evidence of Otter activities come from the distribution of the droppings, or spraints. These oily deposits of fish bones and scales can be seen on particular, and regularly used, grass tussocks and rocks both on and above the shore. The bulk of their diet is probably fish, particularly wrasses and Eels, but they probably also take various types of crab and even Lobsters. It is also suspected that they may also be partial to birds on nesting islands during the summer, and the presence of beheaded adult and juvenile terns on some islands has been attributed to Otters.

There is always the exciting possiblity of finding representatives from the deeper waters of the Lough. In such lowshore areas there may be Common Sea Urchins, Common Stars, and even Curled Octopus and Lobster lying in the pools obscured by the rich kelp growths. Turning over a stone may reveal Eels, brittle stars, some of the larger sea cucumbers, and deeper water sea squirts. In some of the shores experiencing particularly favourable conditions such as strong currents, these species may be able to penetrate even further up the shore. In such places, growths of sponges and sea squirts may cover the rocks, living off the food in the powerful flow of water. Sometimes these sites have rich communities of deeper water animals, including the Light-bulb sea squirt more commonly found at ten metres depth off Ballyhenry Island, and the beautiful Feather Stars, *Antedon bifida*, coloured a rich reddish purple. However, it should be noted that this rich community can be badly damaged by disturbance such as turning over boulders, and in a number of places it is protected by various laws and bye-laws.

Boulders, Grit and Sand

Away from the rock and boulder shores, with their complex weed growths and diverse animal communities, there are other shores equally well endowed with shore life. In many areas where there is moderate wave action, a quite distinctive shore type occurs, of large boulders lying on a relatively flat surface of sands and muddy sands. The distribution of this shore type in the Lough is so widespread that it is easier to state that the only major area where it is not widespread is in the west, to the north of Ringhaddy! The best examples tend to be the beaches derived from recently eroded mixtures of glacial material,places like the south shore around Walshestown, some shores of the Greyabbey area, and the Boretree Islands. In the latter area the erosion is apparently still continuing, providing fresh material for the shore and bed of the Lough. It is possible to speculate that in times past, there must have been many small islands off Kircubbin, until they were eroded, declining into the reefs and pladdies of sands, shingles and boulders that we know today.

Because there are so many types of material on these shores, the plant and animal life is very diverse. The boulders, like their counterparts in the west of the Lough are often covered by weeds, particularly the Serrated, Bladder, and Knotted Wracks. The green, rather spongy branches of *Codium tormentosum* can often be seen. It can be found in winter, but its best growth is in summer. It is particularly abundant out on the pladdies and small islands off Kircubbin, and I always find its deep green a welcome relief to the constant browns of most of the algae growing in these areas. The Sea Lettuce *Ulva lactuca* can also be found in slightly more sheltered situations. In some places the boulders lack weed growth, and there is some evidence to suggest that these patches of bare boulders are increasing in number and extent. Usually the weed is replaced by extensive barnacle populations, and this is often of the Australian species *Elminius modestus*, which was introduced into British waters on ship hulls. Whether the weed decline in these areas is due to the colonization by barnacles, or whether the latter are merely exploiting the habitat released by the disappearance of the weed is unclear.

Boulder, sand, and muddy shore species can be found together in these areas. Shore Mussels form dense clumps on many lower shore boulders, attached by a mesh of remarkably strong threads (the beard, or byssus) secreted by a gland within the animal. The mussels often can be found living alongside Limpets, chitons and Grey Top Shells *Gibbula cineraria*. On some boulders large numbers of saddle oysters (usually *Anomia ephippium*) may occur. These rather inconspicuous molluscs also grow attached to the rock by byssus threads, this time heavily impregnated with lime to form a sort of plug around which the shell grows, in large specimens reaching about three centimetres in diameter.

Many animals that possess shells or hard coverings of some type find themselves colonized by other species. Crabs, with their hard carapaces are often selected by barnacles for settlement, although the fact that the crab moults this external skeleton at regular intervals precludes any long term accumulation of animal life. Shore Mussels are often colonized in this way, and here it can be a much more permanent arrangement, lasting the life of the mussel, which may be as much as ten years. A good example of this is on the mussels growing on the low shore rocks in Greyabbey Bay, where they are virtually smothered by dense colonies of barnacles, small hydroids, and calcareous worms. Even the larger barnacles themselves may be subject to attempts at settlement by young larvae (spat) of both other barnacles and mussels. Perhaps the best local example of this type of cohabitation is from the subtidal levels; a Common Hermit Crab occupying an empty snail shell on which the rather attractive red spotted anenome *Adamsia palliata* is growing. The crab probably derives some protection from this unlikely hitch-hiker, whilst the anenome certainly benefits from the supply of scraps scattered by the crab's scavenging for food.

Under the boulders, and in sheltered clefts, many of the boulder shore species occur in abundance, but this time with burrowing forms in the sediments immediately under the boulders. It is possible to see the thin tentacles of the Strawberry Worm (*Amphitrite johnstoni*) stretched over the interface between mud and boulder, ready to trap surface deposits and small animals moving in the water film. Out on the sands and muds there are many species that we will encounter again on the mudflats, but here one species merits particular attention. This is the Sand Mason Worm (*Lanice conchilega*), quite frequently found low down on the boulder and sand shores in the Lough. Detectable as a scatter of small tubes projecting above the sediment surface, each topped by a fringe of tiny twig-like projections, their tubes extend about eight inches into the sediment. They have a dual life style; as the tide comes in, they extend a mass of tentacles from their tubes, and these can either be used as a filter feeding mechanism trapping food drifting in the water, or for deposit-feeding, by picking up particles from the sediment surface with the tentacles. The tube itself is worthy of examination, for it is constructed of mucus with pieces of shell debris and sand grains, that may give the delicate worm some measure of protection.

Having two feeding methods is not unusual; for many other species, the distinction between scavenging and predation is a dubious one. Crabs and squat lobsters readily take dead and dying animals whilst also actively hunting for shellfish and the more delicate worms. Ragworms (*Nereis virens*), equipped with large and powerful jaws can attack small animals as well as consuming plant debris.

Sands and Muds

Finally, we come to the mudflats. These are extensive areas of sands and muds occurring in many areas round the Lough, but particularly in the north where they spread in an arc from Greyabbey on the eastern side, to Mahee Island in the west. They are at their most extensive at the northern end, between Comber and Newtownards, where over fourteen square kilometres lie exposed during each low tide.

The nature of the mudflats varies enormously. Some areas are com-

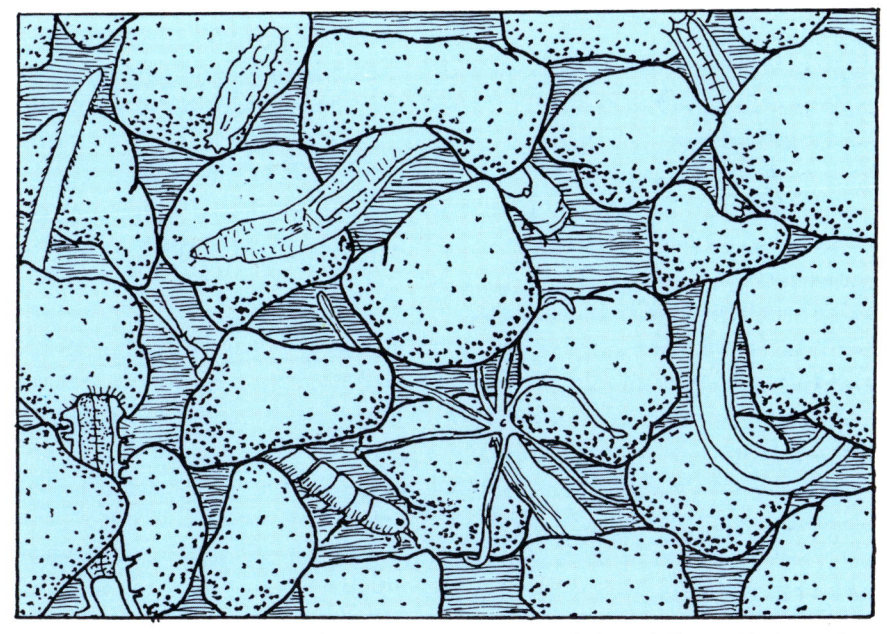

Diagram of meiofauna in sand with a grain diameter of about 0.5mm. Clockwise from bottom left: a gastrotrich, ciliate, annelid worm, turbellarian, nematode worm, hydroid, and a copepod. Such microscopic animals are fundamental to mudflat ecology. (Redrawn from Boaden & Seed, 1985).

posed of relatively firm sands, easy to walk over, whilst other areas nearby can be extremely soft, and treacherous to the unwary. The plant and animal life on these shores varies in similar fashion, according to the nature of the sediments. Superficially, the mudflats in the sheltered bays on the west of the lough for example, may look similar to those off Mount Stewart in the east, but the animal populations are very different in the two localities.

Mudflats are typical of shores with low levels of water movement which have a ready supply of fine particulate matter. In the case of Strangford Lough most of this is derived from glacial sands and silts, the finest materials only settling in the most sheltered corners. Such areas are also the places where organic matter is deposited. Material from the other areas of the lough, the salt marshes, and in some places, carried in by rivers and streams from the Lough's hinterland, finally comes to settle on the mudflats, whilst those same rivers and streams may carry a load of dissolved nutrients as well, encouraging further organic growth. This enormous input of organic matter allows the development of populations of plants and animals frequently exceeding those to be found in the more diverse, but less productive rock and boulder shores of the lough.

Much of this wildlife, like the plankton in the water, is usually never seen. The most widespread plants are microscopic diatoms that grow over the surface of the sediment, often giving it a greenish-brown colour. The most numerous component of the animal populations are the countless microscopic animals, the meiofauna living within the sediments, in the spaces between the individual grains of sand or mud. In the water film between these particles we can find the full range of lifestyles that occur in other parts of the Lough's marine life. Virtually all the major types of invertebrates have representatives here. Thus there are Nematodes, microscopic eel worms, in places exceeding two million to the square metre, and copepods, which can occur in enormous numbers. Many of these tiny organisms, including a number of single-celled species, rely extensively on grazing the bacteria or diatoms or on consuming the particles of organic detritus settling into the sediments. Others are predators or scavengers.

The precise nature of this microscopic world varies considerably according to the nature of the sediments, and the extent to which they allow oxygen and organic matter to permeate down into the water between the particles. Coarse sands, like those found in Millquarter Bay near Killard Point, where there is a relatively high degree of water movement, allow oxygen to penetrate quite deeply into the sediment, but prevent sufficient settlement of organic matter. Such areas may not be so well endowed with meiofauna. At the opposite extreme, the

Eel-grass (Zostera) *growing on mudflats near Castle Espie, with several Periwinkles. Eel-grass is the preferred food for Pale Bellied Brent Geese arriving in the Lough each autumn.*

extremely fine silts to be found in places like Ballymorran Bay or the 'Blind Sound' between Ringhaddy and Whiterock are well supplied with fine organic matter, but the matrix of particles and water may be so tightly knit as to prevent adequate flow of oxygen. Often such areas have a layer of smelly black mud just beneath the surface, revealing the presence of sediments starved of oxygen.

By contrast, the muddy sands found in most of the major mudflat areas in the north of the Lough allow both penetration of oxygen into the sediments, and the deposition of organic matter and nutrients, supporting many of the larger types of plants and animals on the mudflats. The microscopic life in all these areas is essential to the character, and quality of the Lough's waters, for the whole system acts as a giant filter bed, absorbing organic byproducts from the marine life and helping to break down the wastes derived from human activities on farmland and in towns like Comber and Newtownards.

This rich supply of basic organic matter is essential to the growth of the extensive eel-grass swards each summer in the north end of the Lough, and in many of the bays of muddy sands on both the eastern and western shores. They are in fact a group of four different species, but they share common characteristics, being true flowering plants with well developed root systems, essentially similar to those normally found on land. The mudflats start bare in the spring, but from June to October eel-grass pastures develop, giving a green tint to the northern mudflats, and places like the shores round Greyabbey, Ardmillan Bay, and as far south as Gransha Point. The smaller species (*Zostera noltii, Z. angustifolia, and a few Ruppia maritima*) dominate the intertidal areas, but at low water it is possible to find the larger species *Z. marina* growing on the shore and in the shallow subtidal areas. (A good example can be seen during low water in Castleward Bay, but unfortunately it is being severely damaged by the moorings there). This was formerly the dominant species in the Lough, but during the 1930's what was thought to be some type of disease spread across the *Zostera* pastures of all Europe. On Strangford it disappeared almost completely, and the other eel-grass species took over.

Eel-grasses play a fundamental role in the Lough. Their root systems bind the finer sediments, encouraging the accretion of mudflats and even in the later stages, the development of saltmarshes. They increase the amount of organic matter settling on the sediments. They also protect the life on the sediment surface from extremes of heat or cold. The disappearance of these plants in the 1930's may well have had a profound effect on the nature of the lough's sediments, releasing the finer particles, resulting in coarser composition of the sedimentary shores. They are a prime attraction to overwintering wildfowl. We will look at them later; at this stage, however, it is sufficient to stress the

Lugworm burrows and casts at Millquarter Bay.

On the Shore

importance of eel-grasses to the birds (some wildfowl populations declined sharply for a while as a result of food shortage caused by the disease), and the fact that their grazing rapidly depletes the surface growths, so that by the end of November the mudflats have resumed their grey-brown colour, and the root systems are left to recover and regrow the following summer. The birds are not the only agents at work, however. Growth of these grasses seems to vary considerably according to the warmth and rainfall during the summer, and in the autumn, much is removed by gales, usually to be deposited on the strandlines between Newtownards and Mount Stewart, along with other types of detritus.

The most abundant plants of the mudflats are the *Enteromorpha* species - the name literally meaning 'gut shape'. This rather shapeless alga, with its soft tubes filled with air, can be found almost anywhere in sheltered areas, and is often the first species to colonise an area cleared by storms, for example. It is tolerant of fresh water, and because it thrives sometimes to excess in nutrient-rich conditions, it can be an indicator of shores receiving a run-off conveyed by farmland streams, and if growth is excessive subsequent decay may kill off other shore inhabitants. Many bays in the west of the lough have more reasonable quantities of it, and this often attracts wildfowl into such areas. A good example is near the Black Causeway to the south west of Strangford, where in winter Wigeon (*Anas penelope*), Mallard (*Anas platyrhynchos*), and Pale Bellied Brent Geese (*Branta bernicla hrota*) dabble for this nutritious plant.

The mudflats provide a more homogenous habitat than many others in the Lough. Because the majority of animals are burrowing species, there is often an impression that mudflats are barren, wasted, ground. Nothing could be further from the truth, as the winter activities of waders reveal. Even a quick glance across the mud will reveal literally millions of worm casts thrown up by the feeding activities of Lugworms (*Arenicola marina*), each in their individual 'U' shaped burrows, ingesting sand, digesting the organic material and meiofauna within, before expelling the sand at the other end of the burrow. Cockles (*Cerastoderma edule*) occur in dense populations all round Greyabbey, Mount Stewart, and to some extent towards the north of the Lough. In late summer it is possible to find hundreds of tiny young Cockles freshly settled out from their planktonic stage, merely by brushing the surface of wet sand with your fingers. In some areas they may be accompanied by the Sand Gaper (*Mya arenaria*). These molluscs extend siphons above the mud surface to draw in water containing plankton. The siphons of the Baltic Tellin (*Macoma balthica*) are remarkably elongated, allowing them to literally 'vacuum' up deposits on the sediments around them. In many areas the tiny Laver-spire shells, or Mud Snails (*Hydrobia ulvae*) number thousands to the square metre; on one occasion, the sands immediately to the south of Castle Espie actually took on a greyish appearance due to the dense patches of these tiny animals. These work their way through the sediments, feeding on algae, detritus, and even their own faeces, supplemented by bacterial growths. Small crustaceans like *Corophium volutator* also break down this matter, using respiratory currents to draw food into its burrow. In a few localities in the south of the Lough, where sands occur close to a good water supply, it is possible to find Heart Urchins (*Echinocardium caudatum*), a mud burrowing species living a few centimetres below the mud surface. Unlike other sea urchins, which have their mouths in the centre of their underside, Heart Urchins are adapted to burrowing along through the mud, with their mouths to the 'front', for ingesting the sediment, and the spines swept backwards to allow for easier forward movement. All these deposit feeding species play an important role in working the sediments, and allowing oxygen to penetrate, much in the same way that earth worms benefit the soil on land.

The fish of the mudflats are well camouflaged. In the shallow pools left by the receding tide, large numbers of grey-brown gobies lie half buried

CONSERVATION OF THE LOUGH'S SHORES

Shores are prone to reclamation, pollution, dumping, and commercial extraction of materials and food or bait species. Strangford Lough's shores have experienced all of these. Contrary to popular opinion, shores in Britain and Ireland are owned in just the way that land areas are owned; in the case of Strangford Lough, few of these are owned privately; the majority are controlled by the National Trust through ownership and, together with Ards Borough Council, through leases from the Crown Estate Commissioners. The Department of the Environment also own and lease several areas.

All parties try to prevent these problems. However, the frequent occurence of effluent pollution in farmland streams, as well as flytipping in many places where roads come close to the shore has degraded some areas, although the majority are generally in very good condition. Illegal reclamation has also damaged some areas. The greatest wildlife threat to the quality of shore life is that of *Spartina*, or Cord Grass and the National Trust, D.o.E., and Ards Borough Council have all attempted to control this alien through spraying programmmes, so far with only limited success.

Most of the Lough shores have been designated Areas of Special Scientific Interest, whilst a number are also National Nature Reserves. Much of the managment undertaken by the National Trust is in relation to the conservation of overwintering birds. For this, see page 163.

in the sand. Disturb them, and it is only a few moments before they wriggle back into the sediment again. Flatfish like Dab, Plaice and Flounder move over the mudflats to feed on the invertebrates during high tide, taking the exposed ends of Lugworms and enormous numbers of Cockles; one individual that I investigated contained the remains of about ninety after its high tide visit. It is likely that the northern mudflats play a major role as a nursery ground for these species in the lough.

We will be returning to the mudflats again in the succeeding chapters. Their importance for over wintering birds has already been referred to, but their relationship with salt marshes, and the slow accretion of sediments in the most sheltered areas to allow the marshes to become established is particularly important. Moreover, much of the wildlife to be found on the upper reaches of these and all the other shores of the Lough exerts an effect well beyond the limits suggested by the tide line.

6 Above the Shore

THE INFLUENCE of the Lough penetrates well inland. It freshens summer breezes, dulls the edge of winter frosts, whilst salt stains on windows are reminders of gales whipping up the Lough's waters. Nowhere in County Down is very far from some part of the Irish Sea, but the landscapes of the Ards Peninsula, Lecale, and the complex of tiny islands, headlands and bays off the western shore have a particularly intimate relationship with the Lough.

Close to the shore and on the islands, the influence of the sea is particularly strong, with exposure to severe weather being a significant factor. These areas are subject to spray, inundation by spring tides and storm driven waves, and drought. Here conditions are rigorous even for those species adapted for life in such areas. Seaweed, and other marine debris are thrown some way above the high water level, where much of it gradually rots and provides a salty contribution, rich in minerals, to the humus of Lough-side soils. Sediments thrown up by storms provide virgin ground available for colonization by terrestrial animals and plants, but at the same time, other areas may be subject to erosion, with both the wildlife, and the sediments in which it lives, being washed away. Birds also play a major role in such areas by feeding and nesting, and depositing droppings rich in nutrients which in many cases, have been derived from life in the Lough's waters.

Thus a number of the marginal habitats immediately above the Lough's shores provide a challenge to wildlife which is not so frequently encountered below the waves: that of transience, in which pioneering plants and animals must develop and exploit new ground quickly, or alternatively develop a natural succession of different species which, with time, will consolidate the sediments and thus build up a more stable community. By contrast, in other rocky areas of the Lough, there are more stable habitats which can support long lived species clinging to all the clefts and crannies in the bedrocks. They too can have their problems, for these are often in the most exposed areas, subject to the fiercest conditions of wind, wave and drought.

This fringe of land just above the shore is one of the most enchanting features of the lough. Viewed individually, many of the plants are inconspicuous, although some like the Yellow Flag Iris (*Iris pseudacorus*), make up for this with their flamboyance. Together, the wildflower communities impart an almost infinite variety of shades and texture to every curve of the lough shore. As the seasons progress these swards change colour as each species blooms and then fades to be replaced by the next. There are few times of the year when there is not a fresh tint of colour emerging, and this again is a marked contrast to the more constant tones of the shore itself. Close to, even the most inconspicuous plant has its moment of glory, like the Lax-flowered Sea Lavender (*Limonium humile*) breaking out into tiny blue-purple flowers in August, or the star-like flowers of the English Stonecrop (*Sedum anglicum*) poking out of the cracks and folds in the rocks. The animal life of many of these areas is rather less well defined, since it is composed mainly of species coming in from the shores below, the land above, and birds, many of which are long distance travellers.

Strandlines, Stones and Shingle

The most widespread habitat immediately above the shore is that provided by the glacial boulders, shingle and gravels. These are sometimes mixed with boulder clay freshly eroded from drumlin material, but usually the finer elements of this mix are washed away. Large areas of the Lough owe their character to the wildlife of these fringe areas colonised by hardy, salt-tolerant, and usually weatherbeaten plants. Between Newtownards and Kircubbin such areas form an almost continuous fringe along the shore, but they are found in many other parts as well, so that this is often the first true Lough habitat to greet anyone visiting the shore. The plants are easily seen in summer, but in winter many are eroded and beaten back to a few straggly shoots.

The plants and animals of this habitat live on the interface between the marine and terrestrial worlds, and literally have to 'dig in' for survival. They enjoy neither the stability of the slowly accumulating salt marshes, nor the robust foothold of the ancient bedrocks in the south of the Lough. They are the opportunists of the Lough fringe, colonising the fresh material thrown up by the waves or eroded from the land. The loose salty gravel in which they grow means that the only plants to survive are those which are capable of tolerating salty drought in dry weather, fresh water influxes from heavy rain, sudden immersion in salt water during spring tides, and often poor nutrient supply. Such demanding conditions often result in there being patches of bare ground left uncolonised, although in more sheltered areas a dense band of vegetation may develop. This lifestyle is a striking contrast to the relative stability of the lough's subtidal communities where so many species rely on stable,

*Sea Campion (*Silene maritima*) in Castleward Bay, capable of living in thin salty soils on exposed sites.*

predictable conditions, and are adapted to the possiblities offered by having a long life.

Many of the plants are annuals or biennials, producing large quantities of seeds which are often dispersed by the Lough's waters, but only a few of which will get the opportunity to germinate. Perennials are less frequent since survival from one year to the next in this habitat is uncertain, and so their long-term strategies for seed production are unlikely to be successful. Only those species with deep tap roots such as Curled Dock (*Rumex crispus*), or with a dense mat of rooting stems such as Scutch (Couch) Grass (*Agropyron repens*) can hold on during winter storms.

A major feature of these areas is the strandline, that line of accumulated dead weed (and human litter) swept in from the Lough's waters. After autumn gales these deposits can be enormous, particularly on the eastern shores, where they are supplemented by quantities of eel-grass uprooted from the mudflats by the waves. There may be several such lines, reflecting different tide levels in the previous weeks, and a brief glance at these strandlines is sufficient to confirm the difficulties of determining the precise point at which the seashore ends and land begins. One could survey the shore and define it accurately, but the plants themselves are much more ambiguous! Often the 'lowest' of the terrestrial flowering plants like Sea Aster (*Aster tripolium*) and Annual Sea Blite

(*Sueda maritima*) can be found growing below the 'highest' sea weeds like the Channelled Wrack, or even below *Fucus spiralis*. Some of these terrestrial plants may actually be more dependent on moisture from seawater seepage in the ground around their roots, than are the top-shore sea weeds which remain high and dry on the rocks for most of the time.

Strandline debris is a rich food source for animals. Amphipods (sandhoppers) and insect larvae help break down the weed into tiny fragments. Seaweed flies (kelp flies - *Coelopa frigida*) lay eggs in the rotting deposits, and their larvae grow rapidly, literally burrowing into their food supply. The rotting weed is rich and nutritious. In the past it was used by local farmers extensively as fertiliser, but today its riches are mainly important for shore-side plants, providing them with a major source of nutrients in the thin soils. The quantity of food in the strandline is a lure for flocks of Starlings (*Sturnus vulgaris*), Linnets (*Carduelis cannabina*), Yellowhammers (*Emberiza citrinella*) and various finches forage for grubs, small crustaceans, seeds and farmland grain especially in frosty weather. Pied Wagtails (*Motacilla alba*), Grey Wagtails (*M. cinerea*) and in some places Stonechats (*Saxicola torquata*), chase flies and other insects. Badgers seem to take quantities of amphipods by digging through the weed deposits; faeces in their 'latrines' close to the Lough often contain large numbers of dried, pinkish coloured amphipod shells, along with fragments of crab and other materials. There can also be numbers of shrews (*Sorex minutus*) hunting for these small crustaceans; on occasions they can even be heard squeaking from amongst the stones and banks of dead weed.

The most characteristic plants growing on or immmediately above the high water level are the Sea Plantain (*Plantago maritima*), Sea Aster, and the Lax-flowered Sea Lavender, all growing between the stones, and all regularly inundated by the higher spring tides. Common Scurvey Grass (*Cochlearia officinalis*) also grows at this level, in May bursting into a profusion of tiny white flowers. Later, on shingle spits and bars the Sea Campion (*Silene maritima*) dominates the scene with its large bell-shaped flowers, often accompanied by the Sea Pink (*Armeria maritima*). In a few places with sand or fine gravel, for example at both Killard Point and Ballyquintin Point, one can find the Sea Sandwort (*Honkenya peploides*), a perennial plant well adapted by its low, tightly knit growth to windswept conditions. It has a deep root system, penetrating well down through the coarse sand and shingle to moister levels, and fleshy leaves for water storage. In winter, like the other perennials, it dies back, 'retreating' below ground to avoid the winter storms, and in the following spring it sprouts with new shoots. It is believed its seeds must be scratched, or abraded - probably by gravel - before they will germinate.

Slightly higher, and less prone to spring tide immersions, other plants make their appearance. Grasses like Red Fescue (*Festuca rubra*) often form a dense band of vegetation extending some way from the shore. The Halberd-leaved or Prostrate Orache (*Atriplex prostrata*) can be found almost everywhere round the lough at this level, forming an unkempt fringe of grey-green vegetation some two feet high in summer. Often it is accompanied by the Sea Mayweed *(Matricaria maritima)* This delightful plant reaches its peak in August when island fringes and mainland shores are covered with its daisy-like flowers. It has a rich scent that can be smelt even when passing by in a boat. In more sheltered areas, with a little more soil, it is worth searching out the perennial Sea Spurrey (*Spergularia media*), a lowlying plant with tiny pink and white star-shaped flowers. It nestles between the rocks and stones on islands like Lythe Rock, Rainey, and Darragh, near the western shore of the lough, and on the mainland at Ballyquintin Point and Audleystown. Its close relative, the Lesser Sea Spurrey (*S. marina*) can also be found, for example on Taggart Island.

Salt Marshes

In spite of their attractiveness, many of the areas of maritime vegetation around the Lough are not particularly rich in plant species or biologically productive. Notable exceptions are the salt marshes; dense commu-

Common Scurvey Grass growing at the spring tide level.

*Glasswort (*Salicornia*) growing near Castle Espie. The presence of this species is a good indicator that a salt marsh is developing.*

nities of plants growing and spreading over sheltered mudflat areas, and in doing so literally creating their own habitat.

Salt marshes are rare. The best British examples occur in areas like the Wash and the Ribble. In Northern Ireland the richest and most extensive salt marshes still remaining are to be found in Carlingford Lough, followed by the marshes in the north west of Strangford Lough, particularly those along the Comber River. Everywhere, salt marshes are threatened by man-made developments; land reclamation schemes (in reality, the land is not being reclaimed; it is being extended), drainage schemes, airport developments and tidal barrages. In Belfast Lough they have been developed for industry, and in Lough Foyle and at the north end of Strangford Lough for agriculture. Thus the few that remain are particularly valuable, and although the Lough's saltmarshes are relatively small, in the Northern Ireland context they are very important.

The appearance of the salt marshes is not immediately appealing. They are flat and open, and almost invariably the eye reaches beyond them to the Lough's waters. Bleak and usually windswept, their seaward margin merges with the expanses of intertidal mudflats, whilst on the landward side they are often fringed by unkempt scrub and grasses, or reed beds in the damper areas. In many places they are backed by the wide expanses of farmland developed from former salt marshes. They are often intersected by water channels lined with soft muds; these fill as the tide rises, and during spring tides actually flood out over the marsh itself.

This bleak appearance belies their richness for wildlife. The accumulation of nutrients within the sediments encourages dense plant growth, and many of the species present do not just tolerate the constant seepage of salt water around their root systems, they thrive in it. Over decades, and even centuries, of the marsh's development, successive generations of plants germinate, grow, and die back to give way to others. In doing so they not only contribute to the nutrient store of the whole marsh; they also supply the shores and the subtidal areas of the Lough with many essential nutrients for marine life. This may be particularly important for the northern areas, most distant from the tidal flows of the Narrows

The development of a salt marsh is a long and complex process, beginning with the deposition of fine sediments and organic matter in sheltered conditions. Mudflat species like the eel-grasses, *Enteromorpha*, and even some of the brown algae may help by themselves damping down the effects of water movement. With the gradual build up of sediments, conditions become suitable for the first colonist of the salt marsh community. In most places round the lough this is the Glasswort, or Marsh Samphire (*Salicornia spp.*), but in other sandier spots low spikey tussocks of the Common Salt-marsh Grass *(Puccinellia maritima)* may take the lead. Both may be found poking up amongst *Zostera* patches at the upper end of the shores. Glasswort looks very much like a miniature cactus, having a thick fleshy stalk and only rudimentary leaves. The comparison with a cactus is appropriate, for it is in this way that the plant copes with all the problems of storing fresh water in this salty habitat. The thick stem, bulging with water storing cells helps the plant to survive periods of drought or high salinity, whilst the absence of proper leaves prevents excessive water loss by evaporation.

The presence of these little plants is a good indication that a salt marsh is developing or expanding. In Doctor's Bay near Kircubbin, and in the Dorn at Ardkeen the upper muds turn a beautifully fresh green in the summer months as miniature forests of these short fleshy plants grow and flourish. As the developing swards of plants encourage further deposition of mud, often the Glasswort and Common Salt-marsh Grass co-exist, but eventually the latter will take precedence, forming a continuous shaggy sward. At this stage, increasing silt deposition raises the level sufficiently for new arrivals to gain a foothold, and contribute to the developing marsh. These are often evergreen perennials, so they provide the saltmarsh with a protective cover of vegetation throughout the year, helping to consolidate the sediments still further, even during winter storms.

In many places the Annual Sea Blite (*Suaeda maritima*) occurs along with Common Salt-marsh Grass. Like the Glasswort it is small and fleshy, with narrow pointed grey-green leaves. Its low lying growth makes it rather inconspicuous during much of the year, but in August it suddenly turns a deep red that dramatically colours the whole of the salt marsh fringe, before it starts to die back for the winter. In the south of the Lough the Sea Purslane (*Halimione portulacoides*) can often be found along the edge of water channels and breaks in the marsh. Its grey-green colour is a sign of another means by which salt marsh plants can conserve water - the leaves are covered by millions of fine white scales, which help to maintain a film of moist air over the leaf surface, and so reduce evaporation. Strangford Lough seems to lie near the northern limit of this plant; it is seldom found to the north of Whiterock or Kircubbin, whilst in Britain it only rarely occurs north of the Scottish border.

As the marsh develops, the sea-born sediments increasingly become supplemented by humus from the death and decay of plants within the sward. The Sea Plantain (*Plantago maritima*) and the Sea Arrow Grass (*Triglochin maritima*) are typical of such areas, often along with Sea Pink, or Thrift (*Armeria maritima*) which can form dense swards of tightly knit plants, whilst the Sea Aster must be one of the most dramatic. At its peak of growth in August it is about two feet high and crowned with masses of blue-purple daisy-like flowers. Enormous tracts of marsh and topshore are coloured by it, particularly on the wider marshes around Castle Espie and the Comber River. Later, the flowers transform into a mass of silvery seed heads, whilst the Lax-flowered Sea-lavender (*Limonium humile*), the last plant to flower each year, opens its tiny flowers, giving a final tint of purple to the marsh. This species is extremely localised in Britain, whilst in Ireland it is widespread, and around Strangford Lough it is abundant on the marshes and in the more sheltered areas of coast.

The oldest and most developed parts of the marsh are usually those closest to the land. In these areas the marsh has grown well above the level of the shore, and run-off of fresh water from the land plays a greater role in supplying the plants with water. These mature areas have their own range of species, such as the grass Red Fescue (*Festuca rubra*) and the Prostate Orache. Some distinctly salt loving species still occur however, for example the Sea Milkwort (*Glaux maritima*) spreading over patches of ground, and opening into tiny pink flowers in early summer, and of course the almost ubiquitous Scurvey Grass. Places with sufficient fresh water seepage may support the saltmarsh rushes (*Juncus maritimus* and *J. gerardii*) and these represent the peak of the development of the marsh. In many spots round the lough, salt marshes inhibit the flow of water off the land, creating boggy areas. Here, fresh water marsh species may

The salt marsh at Castle Espie.
> a) A developing Glasswort zone;
> b) The eroding section of older salt marsh;
> c) Sward of Common Salt Marsh grass, Thrift, Sea Aster, and Lax-flowered Sea-lavender;
> d) Mature area often supporting salt marsh rushes and Common Reeds;

appear, like Common Reed (*Phragmites australis*) and the Yellow Flag Iris. Such areas can easily be seen from the road to the north west of Castle Espie.

Salt marshes attract animal life both from the land and the sea. Dense populations of small shore species like amphipods, snails and tiny crabs are able to thrive in the lower marsh because of the shelter, and the seepage of salt water. Otters and Kingfishers both hunt along the muddy channels, working their way through the reed beds, and also venturing out into the more open waters when the tide is in. One of the fish to attract the Otters may well be Grey Mullet, which forage around the edge of the marsh for invertebrates and plant matter. As with the banks of stranded weed, these areas are attractive to flocks of small birds foraging for invertebrates and the stock of seeds provided by the Sea Asters. During spells of severe weather in Britain, Snow Buntings (*Plectropherax nivalis*) and Twite (*Carduelis flavirostris*) may join them having been driven westwards to milder conditions. Short-eared Owls (*Asio flammeus*) glide silently over the marsh, hunting for small animals, and apparently quite frequently taking Dunlin (*Calidris alpina*). Occasionally Merlins

(*Falco columbarius*) can be seen darting their way along the shores and over the adjacent fields. The landward side of the marsh is important too; Willow (*Phylloscopus trochilus*) and Sedge (*Acrocephalus schoenobaenus*) Warblers nest in the densest stands of reeds, and in late summer enormous numbers pass through on their southward migration to Africa. Thus for all their bleakness to the human eye, there is a wide range of animals and plants for whom salt marshes are truly hospitable.

Development is not always a one way process. Changing wave patterns, sometimes induced by man-made structures, can result in erosion and even the loss of some salt marshes. To the north of the Rough Island causeway at Island Hill, much of the marsh has been eroded into a series of 'mini cliffs' or isolated clumps and patches as are found at Tank Island (Conly's Point). Sometimes the processes of marsh erosion and accretion occur at the same time, with protruding clumps of old eroded inner marsh being surrounded by patches of Glasswort recolonizing bare mud. In other places, either because of erosion, or through some type of truncated development, the innermost areas of the marsh grow directly on the upper shore, so that reed beds and rushes grow right next to the intertidal zone. The shore road immediately to the south of Newtownards runs alongside patches of this rather curious habitat, which must have been more extensive before the sea wall was built.

There is however, one type of salt marsh that, far from being an asset

Spartina, *or Cord Grass in Ardmillan Bay. The spread of this alien plant threatens many mudflats in the north of the Lough.*

to the Lough, poses a considerable threat to many areas of mudflats. *Spartina anglica*, or Cord Grass, is a vigorous hybrid species originally introduced into the Lough as a means of stabilising a causeway in Ardmillan Bay and a shore at Mount Stewart. Within some thirty years this alien plant has spread throughout much of the northern and western mudflats of the Bay as well as establishing dense swards near Ballydrain, and across the Lough at Gransha Point. Elsewhere numerous isolated clumps threaten further expansion of the species in the Lough. The plant completely dominates areas of mudflats, outgrowing *Zostera*, *Enteromorpha*, and virtually all the salt marsh species described so far. *Spartina* is tough and unpalatable to grazing animals except in the first stages of growth. It spreads mainly by underground rhizomes (but also produces seeds) that allow it to grow into an impenetrable sward that at its most vigorous excludes all other flowering plants, and causes the underlying mud to stagnate. Thus although it may provide shelter for Snipe (*Gallinago gallinago*) and other small waders from time to time, it is virtually useless for most animals. Large areas of mudflats are therefore rendered almost useless to marine life and to overwintering birds. Efforts by both The National Trust, Ards Borough Council and the Department of the Environment to control the species by spraying have met with only limited success, and in some areas are inhibited by conflicts with oysterfarming interests because of concerns about release of silt. There are some areas of England where the species is dying back quite dramatically, but there is no sign of this happening on Strangford Lough yet, and so it remains a major threat to the Lough's saltmarshes and mudflats.

Bedrock and Boulders Above the Shore

Of all the different habitats around the Lough, the harshest must be those provided by the exposed and windswept rocky areas, particularly to the south of Kircubbin, on Ballyhenry Island, and on the two headlands guarding the entrance of the Lough. The exposed conditions prevent the accumulation of soils, water is scarce, and for many plants the nutrients neccessary for survival have to be gleaned from sea spray and bird droppings. These plants are remarkable survivors, adapted for clinging on to rock surfaces, using every nook and cranny, and capable of existing without water for long periods. In one respect, the Silurian bedrocks are ideal for them, since the beds of grit and shale have been folded and contorted so tightly that they provide an infinite variety of crevices for the plants to lodge between. But even the relatively smooth surfaces of the boulders are covered in plant life, and in many of these areas it is almost impossible to determine the true colour of any of the rocks without resorting to scraping some of the plants away.

Rocks, brackish pools and salt marsh at Killard Point. Exposure to the sea has made this area rich in lichens.

Near the the high water zone, the most obvious species are the lichens, growing like a thin crust over the rock surfaces. They are a curious group of very slow growing plants in which a fungus lives in close association with an alga. They are noted for their sensitivity to air pollution; few species thrive in urban conditions, but the north west of Britain, receiving comparatively clean Atlantic air, is well endowed with these plants. Although they are seen widely about the Lough, the greatest profusion of lichens is found in the south. This may be because of the slightly acid conditions provided by the siliceous rocks, which are ideal for some species, and because of the ample supply of nutrients from spray and sea bird droppings in the more exposed areas. It is also worth noting that the regular influxes of smoky Belfast air into the north of the Lough during westerly breezes may be responsible for the poorer lichen growth there.

It is well worth visiting one of the exposed rocky headlands just to see the lichens. They form a colourful series of horizontal zones above the shore in much the same way that the seaweeds do on the shore itself. At the bottom end of the range, we have already encountered the black lichen *Verrucaria*, regularly splashed by waves during high tide. Immediately above the two bright yellow-orange species *Xanthoria parietina* and *Calloplaca thallincola* grow in some profusion. They often have numerous cup-shaped structures, slightly darker than the lichen itself, producing spores in their billions to be carried off by the wind to colonise new rock surfaces. At the upper end of their range they mingle with the grey *Lecanora atra*, and the Crab's Eye Lichen (*Ochrolechia parella*) with its pink-grey discs. Above these, forming the most landward lichen zone, is the extraordinary Sea Ivory (*Ramelina siliquosa*), growing as dense tufts of grey-green branches some two inches long. At the entrance of the Lough these grow in such profusion that seen from a distance, they give the rocks an almost mossy appearance.

Competing with the upper lichen species for the available space are a number of tiny plants so well adapted to their precarious existence on the rocks that they sometimes look like a miniature rock garden. The English Stonecrop (*Sedum anglicum*) and the Biting Stonecrop (*S. acre*) grow anywhere they can insert their roots to obtain the slightest trickle of moisture or condensation, although the latter has a preference for slightly more alkaline conditions. Both flower in late summer and the Biting Stonecrop is sometimes known as Wall Pepper - nibble a piece to find out why! In small pockets of sandy soil between folds and cracks in the rocks plants like wild Thyme (*Thymus praecox*) grow, discreetly producing little columns of purple flowers in high summer, and Sea Pink, one of the most frequent plants in these areas. Unlike the close cropped, tight little rosettes found on the salt marshes, amongst the rocks it becomes quite bushy, with the flower heads on long stalks. Although it can be seen in flower at any time between spring and autumn, its period of glory is in May or June, when all the rocks are covered with a mass of pink flowerheads bobbing in the breeze.

All the plants of these rocky areas have independently evolved the best ways of surviving. Some have developed a low dense growth form, crouched out of the worst of the wind, long roots penetrating deep into the rocks anchoring the plant and making the best of the intermittent water supply, with small, or curled and thin tubular leaves to reduce water loss. Others like the stonecrops are fleshy and succulent to store water during periods of drought, much in the same way as the Glasswort does on the salt marshes.

A few yards inland the folds of bedrock have allowed areas of deeper soil to accumulate, each varying in its acidity depending on how much bedrock is present in relation to the more alkaline glacial debris. There are often small pockets of heathland, and the range of plants depends very much on such variations in quality, and on the degree of exposure to the sea. A number of the plants already mentioned feature in these areas, but there are some delightful additions. On headlands around the entrance of the Lough (and on Gun's Island) the bright blue Spring Squill (*Scilla verna*) is frequent on sheltered banks. In some areas there are thickets of Gorse, locally known as Whin, often in areas where the bedrock is close to, or breaks, the surface (hence the word 'whinstone'). It is possible to find the Western Gorse (*Ulex gallii*) which flowers in late summer, in addition to the Common or European gorse (*U. europaeus*) which flowers throughout the year. However, spring is the main flowering period of the latter, when the rugged shores and the nearby hills are enlivened by brilliant splashes of yellow, and the air is rich with its scent.

On more acid soils, low stands of purple Bell Heather (*Erica cinerea*) intermingle with patches of gorse, moulded by the wind into conformity with the surrounding rocks. In August the Devil's-Bit Scabious (*Succisa pratensis*) blooms with its striking, rather globular blue flowers. Black Knapweed (*Centaurea nigra*) dominates many such areas round the lough, its thistle-like flowers a great attraction to Cinnabar (*Callimorpha jacobaeae*) and the Six-spot Burnet (*Zygaena filipendulae*) moths. These are areas rich in food for Common Blue butterflies (*Polyommatus icarus*). In warm July sunshine they fly in seach of Clover (*Trifolium spp.*), bright yellow Bird's Foot Trefoil (*Lotus corniculatus*), and the Kidney Vetch (*Anthyllis vulneraria*). The damper spots are highlighted by plants like Yellow-Flag Iris and Meadow Sweet, and where proper ponds occur, they can be amazingly rich. During a visit one March, such a pool near the east shore was found to contain several hundred spawning frogs.

No account of Strangford's coast would be complete without refer-

Mating Common Blues on Black Knapweed near Audley's Castle.

A Burnet Rose, growing out of the wind on Ballyquintin Point.

ence to the two very different headlands guarding the entrance to the Lough, each sufficiently rich in wildlife to have been designated as a National Nature Reserve. Ballyquintin Point is the more exposed of the two, and its banks of coarse shingle contrast with the generally finer materials and low sand dunes of Killard Point.

The bleak conditions on Ballyquintin result in a peculiar vegetation. Large areas of cobbles are covered in little except lichens, whilst other species clinging to a precarious existence in the thin soils have taken on a stunted form, beaten into conformity with the ground contours by the wind. There are dense thickets of Blackthorn (*Prunus spinosa*) in places, but these are usually less than a 30cm high - a contrast to the tree-sized specimens to be found in the shelter on the western side of the Lough. Burnet Roses (*Rosa pimpinellifolia*) only 15cm high, carpet patches of gravel with their fragrant creamy blossoms. Other plants include species often found on dry stone walls, like Herb Robert (*Geranium robertianum*), Wall Pennywort (*Umbilicus rupestris*), and Wood Sage (*Teucrium scorodonia*). There are also large areas of rock outcrops, enabling small pockets of saltmarsh to develop, protected from the full extent of wave action crashing in on southerly gales.

Stoats (*Mustela erminea*), noted for their love of tiny holes, clefts and passages in stone walls are very common here. Hares (*Lepus timidus* - the Irish subspecies, *hibernicus*) can generally be seen, but the ground is too

An Emperor Moth (Saturnia pavonia) *on Long Island.*

stony for Rabbits (*Oryctolagus cuniculus*) to burrow. Badgers live in the steep bank on the landward side of the Point.

Killard Point, deriving a limited protection from the headlands around Sheeplands and Gun's Island, is richer in its range of wildlife. Its soils, largely derived from glacial morainic deposits, vary greatly and in parts their sandy consistency introduces habitats different to most other areas of the Lough. Over twenty distinct vegetation types have been identified. Large numbers of Rabbits live on the Point, closely cropping much of the vegetation to the benefit of some plant species, but also digging up some areas with their burrowing, thereby creating new ground for fresh colonisation. These freshly dug areas are often covered in a profusion of Houndstongue (*Cynoglossum officinale*). Along with many calcium-loving grassland species, there are a number of orchids; the Bee Orchid (*Ophrys apifera*) and Common Twayblade (*Listera ovata*) occur in moist lime-rich areas, with Pyramidal Orchids (*Anacamptis pyramidalis*) on the drier spots. Common Spotted (*Dactylorhiza fuchsii*) and Frog (*Coeloglossum viride*) Orchids are scattered over the more neutral grasslands, whilst on areas with a greater clay content it is possible to find the Early Marsh Orchid (*Dactylorhiza incarnata*). Where shallow peaty soils overlie bedrock, Green Winged Orchids (*Orchis morio*) grow; Killard Point is the only known site for the species in Northern Ireland. All these species are protected by law, so the rule is look, admire, photograph or draw, but don't ever pick!

Killard Point is also rich in butterflies. A warm summer with persistent southerly breezes may bring migrants like Painted Ladies (*Vanessa cardui*) and Clouded Yellows (*Colias crocea*), but there are also plenty of native species. Enormous numbers of Common Blues can be seen working their way over the grassy banks, whilst the sun-baked cliff on the south side is also a good site for the rather scarce Grayling (*Hipparchia semele*). This butterfly requires some searching for, as it has the frustrating habit of landing and creeping down to disappear in between the blades of grass, and tilting its wings so as to cast no shadow.

Both points are rich in bird life, not least because they represent a first landfall for many birds flying in from the Irish Sea. Perhaps the species most typical of summer is the Wheatear (*Oenanthe oenanthe*), flashing its white rump as it darts between the rock outcrops along the shore. Piles of shell fragments of the Brown-tipped Banded Snails (*Helix nemoralis*) occur where a Song Thrush (*Turdus philomelos*) has used a stone as an anvil to break them open. Rock Pipits (*Anthus petrosus*) and Skylarks (*Alauda arvensis*) breed in large numbers on both points. It is difficult to visit either area without seeing Stonechats, inevitably perched in pairs, on fence posts, or on the topmost branch of a scrubby bush. Sand Martins (*Riparia riparia*), Jackdaws (*Corvus monedula*) and Starlings successively occupy holes in the sandy cliff at Benderg Bay, whilst Fulmars (*Fulmarus glacialis*) nest on grassy ledges further above. In winter large flocks of

Jackdaw Island, clearly showing its raised beach profile. One of the most important nesting islands, its vegetation is strongly modified by bird activity (C.McC.).

Linnets and Starlings take over the area, whilst careful searching may reveal Black Redstarts (*Phoenicurus ochruros*) feeding just above the shores.

Islands of the Lough

There are few enclosed sea areas in the British Isles so well endowed with islands as Strangford Lough. Even the Scottish lochs, for all their contorted coastlines do not have either the number of islands or the variety of island types. A large proportion of Strangford's islands are extremely important for nesting birds, and this aspect of their wildlife is the subject of a later chapter, but there is much to see besides birds, even on the very smallest islands.

Legend has it that there are 365 islands in the lough; the true figure is closer to 120, but it does depend on how one defines an 'island.' Some of them are so small that it is possible to jump over them! The two Sheelahs south of Greyabbey are capped by small tussocks of grass and Sea Aster little more than a metre across, and in winter much of this is swept away leaving only a tangle of small shoots and bare root ends. The North Rock in the Boretree Islands is much the same, though its instability has been compounded by Rabbit burrowing. In the Narrows, the Black Islands off Isle O' Valla, and the islets in Granagh Bay are nearly as small, but their future is more assured since they are composed of bedrock. In spring the Black Islands are particularly beautiful as their golden crowns of Gorse blaze in extravagant contrast to the blue of the Lough's waters. The Granagh Bay islands have tiny pockets of maritime heath and the folds of the rock are rich in Stonecrop, Scurvey Grass and Sea Campion.

The constant movement of shingle spits, with its effect on the vegetation, gives some islands a very distinctive appearance. The lower part of Dunnyneill Island for example, is capped by a dense crown of Nettles (*Urtica dioeca*) and Alexanders (*Smyrnium olusatrum*) which grows well over a metre high in summer, thriving on the enrichment from sea-bird droppings. It also has a low fringe dominated by transitory species like Mayweed and Prostrate Orache, giving the whole island a slightly top heavy appearance. The other half of Dunnyneill is completely different. Where its steep boulder-clay slopes are sheltered from erosion there is a small wood of Sycamore (*Acer pseudoplatanus*) scattered with Elder (*Sambucus nigra*) and the occasional Chestnut (*Aesculus hippocastanum*). The overall impression is one of a tiny copse echoing to the calls of woodland birds; this is unusual for an island little more than 200m across, and after a day afloat it is a refreshing change from the persistent calls of the Herring Gulls (*Larus argentatus*).

Other islands have been substantially modified by the birds that nest

A Bee Orchid growing on Killard Point (B.B.)

Lichens growing on bedrock outcrops along the Narrows' shore.

on them during the early summmer. On Jackdaw Island for example, there is a fringe of unusually lush grasses. Such areas often reflect their earlier use as a tern colony, and the deposition of vast quantities of droppings within a small area. Whilst the terns are present there is little growth, as the ground is trampled flat and hard, but as soon as the birds have gone, and autumn rain soaks the ground, washing the nutrients into the poor soil, the recovery of the grasses is both rapid and spectacular. This, together with the lack of grazing, results in a number of islands having a dense, tangled mat of rank grassland, often built up into large unsteady tussocks separated by hollows and tiny concealed ditches. In the spring the tussocks are often beautifully clustered with Celandine (*Ranunculus ficaria*) interspersed with Silverweed (*Potentilla anserina*).

Such ground is ideal habitat for small mammals exploiting the very mild and usually frost-free conditions in winter. Brown Rats particularly resort to the islands during cold spells, and wildfowlers have reported rafts of well over a hundred swimming out together to an island. The complex maze of tiny passages between the grassy tussocks must be ideal for them. Rabbits occur (often introduced) on a number of the islands, but they are less suited to such a confined existence, and can outbreed an island's capacity to support them. On the Boretree Islands they stripped the vegetation so completely they had to be eradicated. Moorhens love these areas too, running around between the tussocks, and only bothering to fly when startled into action.

All of the larger islands of the Lough have been farmed for many centuries, and their appearance and wildlife owes much to this aspect of their history. They have remains of old farm buildings, field boundaries of tumbled dry stone walls, and neglected hawthorn hedges still run across their contours. When the low winter sun glances over their pastures, the undulations of the old potato cultivation ridges ('lazy beds') can still be seen running up the island slopes. In spite of the liming, fertilizing, and drainage that must have been applied to the islands by those hardy farmers, many retain a significant wildlife interest since their inaccessibility has helped to safeguard them from modern farming techniques. Thus we still can find some rich, and fairly natural meadows on a number of the islands, small thickets of blackthorn, ash, and even occasionally Spindle trees (*Euonymus europaeus*), in autumn, decked with clusters of bright pink fruits which split to reveal four orange seeds. They often reflect the presence of a limey, or alkaline, component to the glacial till. Pools, originally dug for cattle, now harbour their own wildlife, and are fringed with rushes and Yellow-Flag Iris. On Salt Island there is a thriving population of Irish Hares.

One of the most attractive islands is Darragh, lying to the north of

Ringhaddy Sound, and passed by countless boats making their passage from Whiterock to the southern parts of the Lough and beyond. In May and June the deep mauve spikes of the Early Purple Orchids (*Orchis mascula*) poke up through the vegetation, competing with Violets (*Viola riviniana*) for attention. Later, in the damper areas, the Common Spotted Orchids bloom in an amazing range of colour varieties from pale purple speckled with red through to an almost pure creamy white. They are often mingled with the dense stands of Yellow-Flag Iris, anything up to three feet high. The meadows on the island are a mass of flowers and delicate grassheads waving in the breeze. In drier parts the Bell Heather can be seen, along with the bobbing blue heads of the Devil's Bit Scabious. The latter is one of the food plants of the very localized Marsh Fritillary (*Euphydryas aurinia*). It is not a large butterfly, but its marbled pattern and rich colours make it an exciting find.

The butterfly life of the other islands is full of variety (within the context of Ulster's limited butterfly fauna). Mild spring days nudge Speckled Woods (*Pararge aegeria*) into early action in the shelter of thickets of scrub, where they can be seen competing for territory and sunny spots. Damp areas with Ladies Smock (*Cardamine pratensis*) have their attendant Orange Tips (*Anthocharis cardamines*). The derelict stone walls often attract the appropriately named Wall Brown (*Lasiommata megera*), sunning itself in the warmest corners. High summer often sees large numbers of Common Blues and Small Coppers (*Lycaeana phlaeas*) and Taggart Island can be amazingly rich in these two species, particularly along the sunlit western side of the island in mid afternoon. There they are usually accompanied by Meadow Browns (*Maniola jurtina*) and Ringlets (*Aphantopus hyperantus*) flitting about in the long grass.

Behind the Coastal Strip

Inland from the Lough's shores, the more characteristic mainland scenery of County Down begins to take over. Both the mainland and a number of the larger islands have been inhabited for about seven thousand years, when the first mesolithic hunters established camps on the raised beaches near the shore. Thus there are very few areas where the landscape has not been modified by man's activities. In spite of this, the influence of the Lough on the wildlife and its habitats is still clearly detectable, and its richness for wildlife is reflected in the bird populations. Of the 155 species of bird recorded in Northern Ireland during the surveys for the B.T.O./I.W.C. Winter Bird Atlas, 122 or 79% were present on Strangford Lough and its hinterland.

The original woodlands of Oak, Ash, Willows, Alder, Hazel and Elm, that developed about the Lough in post-glacial times must have provided

A young Fox on Darragh Island (D.A.).

a glorious setting for the Lough's waters. There are still one or two patches of woodland that may be quite close to their original counterparts. The old oak wood on Rainey Island near Whiterock is a good example, with its trees hanging over a rich shoreline. In spring it is particularly beautiful, with drifts of Bluebells (*Hyacinthoides non-scriptus*) growing in the shade, and always the chance of seeing the Long-eared Owls (*Asio otus*) that regularly haunt the wood. Skillins Wood near Greyabbey is also well worth visiting; it is on a small headland overlooking Greyabbey Bay, and the trees have been moulded and contorted by the wind around its edges so that a protective belt is formed for the trees within. This is quite a damp site, and there are numerous Willows and Alders. However, most of the woodlands round today's lough (increasingly frequented by Buzzards - *Buteo buteo*), in places like Castle Ward, Delamont, Finnebrogue, the Nugent Woods at Portaferry, and Mount Stewart, are all products of planting schemes undertaken in the eighteenth and nineteenth centuries. Were it not for these schemes, the Lough would be bereft of any sizeable areas of woodland, but it has involved the reintroduction of Scots Pine (*Pinus sylvestris*) and the introduction of alien species like Beech (*Fagus sylvatica*) and Black Pine. Sycamore is another alien, introduced in mediaeval times, and now it is second only to Ash as the most ubiquitous tree in our hedgerows.

These areas of woodland are very much a minority feature of the landscape. Most of the land round the Lough is given over to agriculture of different types. To the north, the slopes of Scrabo Hill are covered in large open fields, mainly devoted to cereal production. This has intensified in recent years, with the attendant destruction of hedgerows and other field boundaries. Below Scrabo Hill, on the reclaimed land protected from the Lough's waters by the tidal bank, lies some of the most productive farmland in the Province. Flat and open, intersected by a series of drainage ditches, the large, prosperous farms are mainly devoted to market gardening, with Belfast as the main outlet for the produce. The land around the south of the Lough, though once regarded as the 'Granary of County Down,' is now generally regarded as being of poorer quality. This may be so, but what it lacks in commercial value is generously compensated for by scenic beauty, with its small rugged farms, tiny fields with lichen-encrusted dry stone walls, and Gorse covered rock outcrops. Between these two extremes lie the majority of the Lough's farmlands; small farmsteads tucked between the rolling drumlin scenery, with well-fed cattle thriving on almost unbelievably green pastures, fringed by dense hedges of Hawthorn (*Crategus monogyra*), Brambles (*Rubus fruticosus*), Holly (*Ilex aquifolium*), Sycamore and Ash. In terms of their wildlife, many of these areas have experienced the same 'improvements' as other agricultural landscapes. Virtually all the grassland has been reseeded with more productive grass mixtures, and this

A Small Heath (Coenonympha pamphilus) *on Slievenagriddle, near Saul.*

National Trust sign in Castleward.

together with the application of herbicides has resulted in the widespread disappearance of meadows and their rich array of wildflowers. The switch from haymaking to silage production, resulting in earlier cutting each year, and increased use of fertilizers may have been one of the factors responsible for the virtual disappearance of the Corncrake (*Crex crex*) from the County. Drainage or infilling of most of the damp hollows has also resulted in a loss of important wildlife habitat. Nonetheless, compared with the agricultural wastelands of areas like East Anglia, the farmlands around the Lough are full of interest. For example, large flocks of birds like Redwings (*Turdus iliacus*), Fieldfares (*T. pilaris*), Yellowhammers, Finches and Tree Sparrows (*Passer montanus*) can be seen working their way along the dense hedgerows in winter. These hedgerows also provide 'corridors' for species like Badgers and Foxes, whilst dry stone walls are an ideal habitat for Stoats. In many areas, particularly in the south of the Lough, there are still large numbers of Irish Hares. There are also a number of landscape features that are not amenable to agriculture, and therefore even 'well managed' farmland has retained its interest in places.

One of these features is the number of steep drumlin banks eroded by wave action during the periods of higher sea level after the Great Ice Age. They have remained uncultivated, and apart from occasional burning, free from direct human influence. They frequently occur immediately inland of the shore road where it runs along the raised beach, and on headlands looking over the eastern shore (a good example is the bank behind Kircubbin Sailing Club). Often they are covered in impenetrable thickets of Gorse, Blackthorn, Hawthorn, Ash and Sycamore. These are ideal places for Foxes and Badgers; often a careful search will reveal ⟨ in the dense undergrowth. In areas of open pasture, they provide cen for the small birds of the area. Where the thickets overlook the shore it is often possible to find Stonechats and Reed Buntings (*Emberiza schoeniclus*), whilst the long grasses and tangles of Brambles may hold wildfowl nests.

Arguably one of the most important natural features still regularly seen on farmland around the Lough are the small lakes and marshy areas lying between the drumlins. Their origin also dates back to the periods of higher sea level, when the shores of the Lough must have been even more contorted than they are today. Sheltered conditions in the interdrumlin hollows allowed the deposition of fine impervious clay. As

---

CONSERVATION OF SOME TERRESTRIAL HABITATS

There are a number of habitats that are particularly important to Strangford Lough, and have managed to survive either because of their isolation on islands or other remote areas, or simply because they have not been developed. Many aspects of these are important to the Lough's wildlife as well as to its scenery.

• SALTMARSHES   Several of these have been aquired by the D.o.E. and the National Trust in order to preserve this rare habitat from reclamation and dumping, which have badly damaged marshes elsewhere.

• NATURAL MEADOWS   Most of these have been lost on the mainland through reseeding, fertilization and drainage. On those that remain, regular grazing in winter, with an absence of artificial fertilization, allows a rich array of wildflowers to develop in spring and summer, supporting abundant insect life.

• SCRUBLAND   Areas of Blackthorn, Bramble and Elder thicket provide pockets of shelter on islands and other exposed sites, and are important for the smaller birds for nesting and feeding. These are therefore encouraged in some areas, but controlled in others so they do not colonize or exclude other habitats.

• WOODLAND   Much of the Lough's woodland was planted in the 18th and 19th centuries, and many trees are now mature or even old and dying. Often there has been no natural regeneration, and no replanting. Removal of some of the old and decaying trees creates clearings for replanting with native species, whilst others, left to die and decay naturally, can support a wide range of wildlife.

the sea level dropped these hollows were separated from the sea, and became natural collecting areas for water running off the drumlins, thereby forming numerous small pools and lakes. Plant and animal life immediately colonised these areas. Reedbeds began to grow in from the fringe, and with the accumulation of plant matter and silt, marshland was created round the edges of the ponds, often still containing reeds, but also holding moisture-loving trees like Willow and Alder. In some cases, the marsh has grown completely over the lake, and the accumulations of dead plant material, poor in oxygen, instead of rotting, have formed raised peat bogs. Others, possibly with steeper shores inhibiting the spread of the reedbeds, still hold stretches of open water.

These areas of water and wetland are immensely rich in wildlife. The acid conditions of the raised bogs led to the formation of peat from *Sphagnum* moss. This was almost invariably cut for fuel, leaving bogs with dry heather ramparts and wet hollows full of Bog Cotton (*Eriophorum angustifolium*). Bracken and Devil's Bit Scabious often colonize the ramparts. Such places include Inishargy Bog near Kircubbin, and Turmennan near Downpatrick. The marshlands, in addition to the reedbeds, hold a large variety of sedges, rushes, and plants like the Marsh Cinquefoil (*Potentilla palustris*), Bog Bean (*Menyanthes trifoliata*), and Marsh St. John's Wort (*Hypericum elodes*), all thriving in the damp conditions, whilst the lonelier stretches of water still support large numbers of Otters.

Gorse, or Whin, growing at the Black Islands, near Strangford (A.J.).

Above the Shore

Sea Mayweed on Jackdaw Island. Weed and debris tossed up by storms rots and provides a rich supply of nutrients and minerals for growth.

For Strangford Lough, the greatest importance of these wet areas is the role they play in the bird populations, particularly in winter. This is the subject of a later chapter; here it is sufficient to note that these lakes and marshes provide excellent winter feeding for Greylag Geese (*Anser anser*), Mute Swans (*Cygnus olor*), Mallard (*Anas platyrhynchos*), Teal (*Anas Crecca*), and Snipe. In many areas there are regular flightways between the Lough and places like the Clea Lakes, Carrigullian Lough and Lough Cowey, and the birds may resort to these in responding to changes in the tides, weather, or food availability. Of all the different aspects of the Lough's bird life that have been examined, this is the one we know least about, but it is known that these small lakes play an immensely important role for some of the bird populations, and indeed wild Greylag nested in such areas a hundred years ago.

One area of fresh water merits particular mention. This is the Quoile Pondage, formerly the estuary of the River Quoile opening into Strangford Lough. Because of regular and severe flooding in lowlying Downpatrick, a barrage was constructed linking Castle Island and Hare Island to each other and to the mainland on each shore of the estuary. By providing an alternative holding area for flood waters during high tides, this prevented them from backing up the river, thereby avoiding the flooding caused by the combined effects of heavy rain and spring tides. Drainage of the new basin was catered for by a series of sluice gates which open as the tide falls and allow the river water to flow out during low tide, closing again as the tide rises once more.

The construction of the barrage in 1957 fundamentally changed the nature of the whole area. Originally, extremely sheltered estuarine conditions prevailed, backed by an extensive area of regularly flooded wetlands - the Downpatrick Marshes, themselves created by an earlier scheme. Since then, the waters of the Quoile Estuary have become almost fresh (though with a slight brackish element due to seepage), whilst the marshes have become much drier. Despite the barrage, a few Grey Mullet, Sea Trout and Salmon slip in through the sluices, whilst Eels head out to sea to breed. The seaweeds and marine shore life have been replaced by fringes of reeds and other freshwater grasses and sedges, Willow and Alder on the lower shore, with Oak and Ash on higher ground. The open stretches of water now support swans, Coot (*Fulica atra*), Great Crested Grebe (*Podiceps cristatus*), and in the summer, flocks of terns feed near the barrage. Grazed flood meadows still exist within the area however, and these provide feeding for Wigeon (*Anas penelope*), Greylag geese coming in from the Lough, and for other fresh water specialists like Pochard (*Aythya ferina*). The processes resulting from the creation of the freshwater pondage are still continuing, and it is possible that the scene in thirty years time will be as different from today as it now is from the former estuarine conditions. The Pondage has thus provided a rare opportunity to observe colonization, and this was recognised by the areas's designation as a National Nature Reserve.

In summary, the range of different wildlife habitats around Strangford Lough's coast that have featured in this chapter, and their many counterparts on the shores and in the depths of the Lough operate as a single system, united by the complex topography of the Lough, by the movement of its waters, and by the interchange of nutrients and energy between the species exploiting the different habitats. Few such coastal areas in Britain or Ireland have anything even approaching the richness and diversity of this system. It is unlikely that we shall ever have a complete understanding of how it operates, but appreciation of this fundamental unity may help us to interpret the way different animals and plants respond to the Lough, and hopefully to look after it.

7 The Wintering Birds of the Shore

SO FAR, we have been looking at Strangford Lough very much as at a stage, in which some three thousand different types of animals and plants play their various roles. It is now time to shift the emphasis, and concentrate on just one small group, the birds, and see how they exploit the Lough's resources. There is no compelling scientific reason for this; in fact in terms of their numbers, or by any other standards of measurement, they form a tiny minority of the wildlife. But they command more attention from the human observers of the Lough than any other group, because they are easily seen, because of their beauty and variety, and because they are often good indicators of the state of the Lough itself. They also, more than any other type of animal, bring an international dimension to the wildlife by virtue of their great mobility. The wintering birds illustrate this particularly well, and their relationship with the Lough makes an exciting story.

With the approach of winter, almost all the birds breeding in the northern parts of the world begin to move south. For coastal birds arriving to spend their winters in Ireland and western Britain, Strangford Lough represents a major feeding and sheltering area. As the winter progresses and coastal birds work their way along the Irish Sea, the Lough provides one of the most important links in a chain of coastal inlets and estuaries. Later in the winter, it is an area of secure feeding as food in other coastal sites becomes depleted, and finally, with the approach of spring, many north-bound birds stop over briefly to feed, refuelling before embarking on the long journey north to their breeding grounds.

The birds start appearing in the Lough as early as July, with the arrival of scattered flocks of Dunlin, Turnstone (*Arenaria interpres*), Bar-tailed Godwit (*Limosa lapponica*), and other waders. These are the birds that have failed to breed successfully, and have left the breeding grounds early. Many are surprisingly tame from their stay in the remote north. It is not however, until October that the populations begin to approach their maximum numbers. From then through to March, the Lough holds important stocks of some fifteen species of wildfowl and about the same number of waders, in addition to several species visiting less regularly,

Wintering Pale Bellied Brent Geese. The Lough has been designated a Sister Reserve with North Bull Dublin, and Polar Bear Pass Canada, because of this important species (W.McA)

and in smaller numbers. Populations of the migrating species frequently exceed the criteria of both international and national importance.

These accumulations of birds are an impressive sight. Flocks of Pale Bellied Brent Geese, often with well over five thousand birds in a single flock, rise and settle again in response to tidal movements. For people living near Strangford Lough, the throaty calls of the Brent echoing across the distant mudflats are hauntingly evocative of autumn. As the tides advance and retreat, flocks of Knot (*Calidris canutus*) circle over the water's edge. Sometimes they are so numerous they can appear to the distant observer more like a swarm of bees as they weave and twist in the air. Occasionally the whole flock seems to change colour as the birds turn in unison, and the feathers of several thousand pale breasts suddenly reflect the low winter sunlight. Below, on the mudflats, other birds work their way along the water's edge, Oystercatchers (*Haematopus ostralegus*), Dunlin, and Curlew (*Numenius arquata*) all hunting for the invertebrates buried beneath the mud. The open expanses of mudflats, sheets of water reflecting the pale sunlight, the melancholy call of the Curlews carried on a chill breeze, all combine to create their own particular beauty. Why is Strangford Lough so important for these birds?

The main reason is the richness and variety of its shoreline habitats, and the food they hold. The mudflats are among the most extensive in

The Wintering Birds of the Shore

Ireland, and are relatively unpolluted and undamaged. Stretching from south of Comber across to Newtownards they provide over fourteen square kilometres of feeding ground, with more large areas to the south in Ardmillan Bay, off Mount Stewart and Greyabbey, and in other parts of the Lough. As we saw in Chapter 5, this habitat is richly stocked with a variety of foods attractive to birds, which after a summer of growth and breeding, are particularly abundant. Eel-grasses, algae, worms, snails, Cockles, and small burrowing shrimps all combine to provide a rich harvest for both wildfowl and waders alike as winter advances.

The shores of bedrock and boulders, with their luxuriant growth of seaweeds, provide an equally rich hunting ground, particularly for waders. On the sheltered western shores, in the small bays to the south of Whiterock like Quarterland and Ballymorran Bays, the soft muds attract their own selection of birds seeking worms and algae in those quiet corners. Many shores have small streams, and these support a range of different food species, and provide fresh water for the birds to wash and preen. The adjacent farmland, with its small interdrumlin lakes and bogs is also important, and together with the salt marshes and small islands, provides an additional selection of areas for the birds to exploit.

In addition to its importance as a feeding resource, the Lough is an area in which birds can find shelter in most weather conditions. This may surprise those who have watched small flocks of birds braving the autumn

Knot over the mudflats near the Comber River estuary. Such displays are important in their daily feeding and roosting cycle (L.J.T.).

gales sweeping in from the south west. However, in contrast to open sea areas, the landlocked Lough is indeed sheltered, and the numerous small islands and pladdies mean that in any weather condition there are going to be some corners to which the birds can retreat. In addition to that, as we saw in Chapter 3, the Lough and its environs are relatively frost free compared with many other parts of Ireland and Britain. However, when some shores do actually freeze over, the effects can be severe, as we shall see later.

Finally, it has to be said that the Lough's present importance for wintering birds is enhanced considerably by the problems and threats in other esturaries and inlets. Many have lost, or suffered damage to, intertidal habitats and salt marshes through development and land reclamation. Strangford Lough is not immune to these problems, but compared with Larne Lough, and the continuing destruction of estuarine habitats in Belfast Lough for example, it is in a relatively natural and undamaged condition. To say that the Lough has attracted birds displaced from these other areas would be an oversimplification, but it does provide a superlative example in a declining number of refueling points essential to thousands of birds in their winter movements through western Europe.

On the Move

Migration is a drastic undertaking for any population of animals. Many of our overwintering birds cover immense distances in their journeys from their breeding grounds, usually over open sea, and at seasons often characterised by equinoctal gales and cold weather. The number of birds that die through storms, starvation, and loss of direction is unknown, but these factors must be so severe that only the very fittest individuals survive.

The origins of these migrations are complex, reaching back to the Great Ice Age and the existence of ice-free refuge areas where the birds could continue to breed. These were located in northern Greenland, around the south of what is now the North Sea, and in Russia. With the arrival of warmer conditions, and the retreat of the ice, the birds extended their breeding ranges from these areas as new land became available.

It is still possible to recognise these old ice-free areas in the modern summer distribution of the birds wintering in Strangford Lough and elsewhere. Knot, Turnstone and Brent probably centered on the Greenland refuge some 18,000 years ago, and now breed in northern Canada and Greenland, the Brent on Bathurst Island qualifying as one of the world's most northerly breeding birds. Whooper Swans (*Cygnus cygnus*),

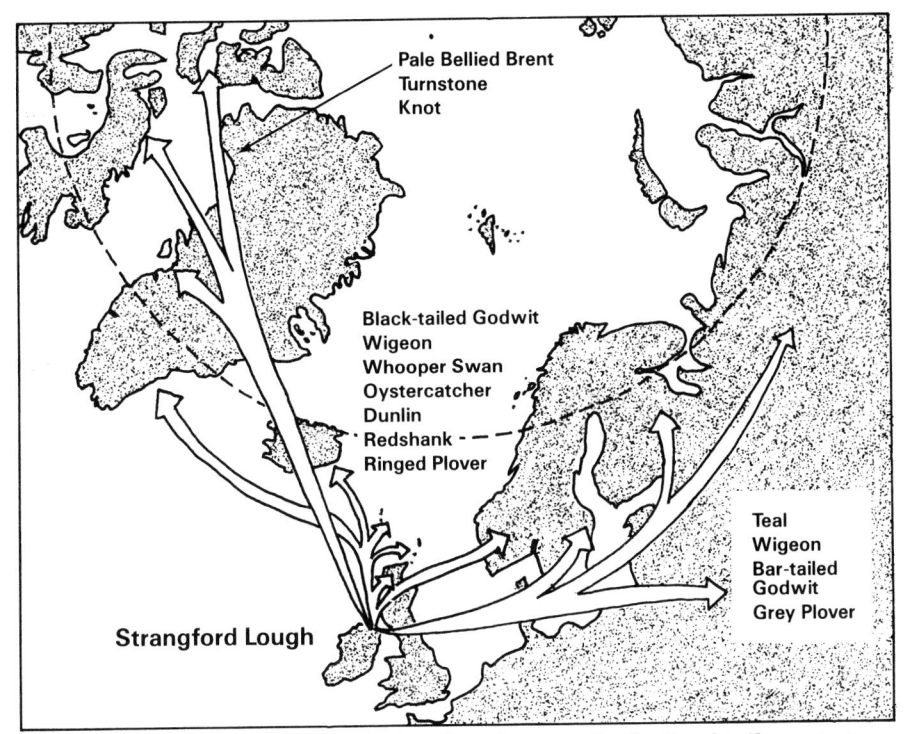

Strangford Lough's overwintering birds; where they nest. For further details, see text.

Oystercatchers, Ringed Plover (*Charadrius hiaticula*), Dunlin and Redshank (*Tringa totanus*) were once confined to the area round what is now the North Sea. They have extended their breeding range to include south Greenland, Iceland, Britain, and Scandinavia. Oystercatchers and Ringed Plover are particularly notable as breeding birds on the Lough. The Russian refuge areas of the last ice age still provide breeding grounds for many of the Lough's Bar-tailed Godwit, Grey Plover (*Pluvialis squatarola*), Teal and Wigeon (these also breed in Iceland).

These migrations stretch the physiological abilities of the birds to their limits. Before departing on the first stage of a migration, it is essential for the birds to feed and put on weight - in other words, fuel - to be able to undertake the journey. Usually this is in the form of fat reserves, which in the case of Knot may be up to 80% of the lean body weight. How successful they are in doing this has a material effect on the likelihood of surviving the journey, during which they may have little opportunity to even land and rest, let alone feed. For example, Knot may lose up to 50% of their body weight during various stages of their migration, and they must restore this (ie, refuel) before resuming the journey.

In spite of the complex reasons behind these arduous journeys it is

Golden Plover on Gull Rock, Mahee Island, bearing summer plumage (L.J.T.).

clear that in the long term there is still an advantage to be gained for the migrating birds. In other words, more individuals survive by leaving their northern breeding grounds and moving south, coming to places like Strangford Lough, than would do if they stayed behind. Although the breeding grounds may be suitable for the birds to nest, rear and feed their young in summer, this period is restricted, and in good breeding years such areas may have to support twice the number of birds feeding later in the season than were present at the beginning. In the case of some waders whose young are less efficient at feeding themselves than their parents, many adults leave the breeding ground early, thus helping to prolong the availability of food for those remaining. Even for our own coastal species there is often a struggle to feed the young at a time when shoreline food species are only just recovering from the intense mortality of the previous winter. Oystercatchers, for example, cut down their own food consumption to give the choicest items to their young, and as a result may lose weight quite considerably at this time.

As the onset of snow in the high arctic heralds the approach of winter, and a succession of depressions batter the coasts and islands of the north Atlantic, the northern breeding grounds soon become inhospitable. Insect life becomes scarce, burrowing worms dig deeper to avoid frosts, and plant growth stops. Food stocks for some species may disappear entirely. So they leave, northern breeding species like the Brent deserting the Arctic at a time when Strangford Lough is still enjoying its summer weather. Indeed, for much of the coming winter on the Lough, temperatures may be equal to, or even warmer than those experienced by the Brent in their Arctic summer. Non-breeding birds, unhindered by inexperienced offspring leave first, followed shortly after by family parties, all flying south east across Greenland, Iceland, and finally at the end of August the first Brent geese cross the coasts of Antrim and Londonderry. Sometimes, on a September evening, they can be seen high in the sky over Newtownards and Comber beginning their descent on to the northern mudflats, finally seeming to tumble almost gratefully out of the sky at the end of their long journey.

Some of the first species to arrive still bear the summer plumage of their northern breeding grounds. A visit to the Lough shores at this time can be especially rewarding with so many birds on passage pausing to feed briefly before moving on. Both Grey and Golden Plovers (*Pluvialis apricaria*) may be seen with the startling black breast plumage of summer, and Bar-tailed Godwits often retain their rich chestnut hues before changing to the drab grey-browns of winter. Perhaps the most colourful are the Turnstones (some non-breeding birds remain in the Lough throughout the summer) with their richly mottled colouration of a very pale grey, black and deep red brown. It is at this stage of the year that the Lough receives sudden influxes of birds on passage to other areas, causing numbers of some species to increase spectacularly for a while, before the true winter residents settle in.

Thus from late summer through into the New Year, the Lough is the setting for a constant series of arrivals and departures, with numbers of birds gradually increasing. Wigeon make their appearance in September shortly after the Brent, the first flocks arriving from Iceland, and later arrivals being birds from eastern Europe and Russia, having worked their way across the continent and Britain. Whooper Swans usually arrive in November, whilst Shelduck (*Tadorna tadorna*) are still arriving well into January.

When the population of each species does reach its peak, the numbers can be very impressive. At the peak of their season (there has been a trend towards earlier arrivals in the last twenty years, and they currently peak in October) about two-thirds of the whole European population of Pale-bellied Brent may be present. Because a number of Brent have been marked with individually coded leg bands, we know that some individuals move on even before numbers reach their peak. The proportion of Brent actually relying on the Lough at some stage or other is therefore considerably greater, possibly as much as 90% of the total European population.

WINTERING WILDFOWL AND WADERS ON STRANGFORD LOUGH

Regular overwintering waterfowl, with the usual peak counts, followed where appropriate, by their significance (based on the internationally accepted criteria of over 1% of either national or international population levels). Those of Ireland should be regarded as tentative, due to shortage of data. Species breeding on the Lough are indicated: -B-

SWANS

Mute Swan	*Cygnus olor*	Over 200; UK; -B-
Whooper Swan	*Cygnus cygnus*	ca. 90; UK

GEESE

Pale-bellied Brent	*Branta bernicla hrota*	Over 13,000; International
Greylag	*Anser anser*	Over 400, mostly feral; -B-
Greenland White-front	*Anser albifrons*	Rare, formerly common.
Barnacle	*Anser leucopsis*	Introduced; over 50; -B-
Canada	*Branta canadiensis*	Introduced; up to 200; -B-

DUCKS

Mallard	*Anas platyrhynchos*	Up to 2000; -B-
Teal	*Anas crecca*	Up to 1000; UK
Wigeon	*Anas penelope*	Large decline; now under 2000
Pintail	*Anas acuta*	Over 200, Ire
Shoveler	*Anas clypeata*	ca 130; Ire & UK
Shelduck	*Tadorna tadorna*	Over 2000; Ire & UK; -B-
Gadwall	*Anas strepera*	Under 100. Ire & UK; -B-
Pochard	*Aythya ferina*	Ca 300, Ire
Tufted	*Aythya fuligula*	Over 300 Ire; -B-
Scaup	*Aythya marila*	Under 30; Ire
Goldeneye	*Bucephala clangula*	Over 400; Ire & UK
Red-breasted Merganser	*Mergus serrator*	Up to 300, Ire & UK; -B-

WADERS

Oystercatcher	*Haematopus ostralegus*	Up to 5000. Ire & UK; -B-
Ringed Plover	*Charadrius hiaticular*	Over 200; Ire; -B-
Golden Plover	*Pluvialis apricaria*	9000 & over. Ir & UK
Grey Plover	*Pluvialis squatarola*	Under 200, Ire
Lapwing	*Vanellus vanellus*	Up to 14000, Ire & UK; -B-
Knot	*Calidris canutus*	10000 & over International
Dunlin	*Calidris alpina*	Up to 6000, Ire & UK
Purple Sandpiper	*Calidris maritima*	Scarce but regular in south.
Common Snipe	*Gallinago gallinago*	Up to 100, prob undercounted
Jack Snipe	*Lymnocryptes minimus*	Scarce, but undercounted.
Black-tailed Godwit	*Limosa limosa*	Up to 200; Ire & UK
Bar-tailed Godwit	*Limosa lapponica*	Over 1000 Ire & UK
Curlew	*Numenius arquata*	Up to 2000, Ire & UK
Redshank	*Tringa totanus*	Over 3000, International; -B-
Greenshank	*Tringa nebularia*	Up to 40, Ire & UK
Turnstone	*Arenaria interpres*	Over 400, Ire

Shelduck sometimes occur in internationally important numbers, whilst other wildfowl species may all occur in numbers regarded as being of national importance. The situation is similar with waders; Redshank and Knot achieve numbers of international importance and the majority of the others reach levels of national status. The precise numbers of some species have fluctuated considerably over the years. In part these changes can be attributed to biological and climatic influences, but in some cases human events have played a role. In particular, a persistent decline of Wigeon numbers since the early seventies has given cause for concern about the effects of human disturbance, particularly wildfowling, on the Lough.

Strangford Lough can also be the scene of further influxes much later in the winter. Because of its north-westerly position in the British Isles, the mild winters allow the Lough to act as a refuge for birds avoiding severe winter conditions in Britain and on the continent. On such occasions enormous numbers of birds may move westwards, considerably swelling numbers of species like Bar-tailed Godwit and Lapwing on the Lough and other places along the Irish sea coasts.

Winter Feeding – It's Importance

The most critical resource for the birds arriving to overwinter in Strangford Lough is food. For all overwintering birds in northern Europe, where conditions can change from mild Autumn sun to subarctic snows within a matter of days, the distinction between adequate feeding and starvation can be a fine one.

Birds, like mammals, are warm blooded; they maintain their body temperatures at a more or less constant level irrespective of the prevailing air or water temperatures. In cold weather they can manage this in part by adjusting their behaviour; look at any flock of birds roosting on a shingle spit round the Lough during a stiff winter's breeze and you will see they all face into the wind, huddled together. Because of their streamlined shape the birds offer less resistance to the wind by doing this and their feathers are not being constantly ruffled by fresh gusts of cold air. They also roost with one leg tucked up into the feathers to reduce heat loss still further. In severe conditions however, the birds have to do more than just that. In order to maintain their internal warmth they have to use up more energy to counter the increased heat loss. That energy can only come from food, or from stored energy originally obtained from feeding in less stressful conditions. Small birds, with a large skin area to their body volume, use up relatively more energy than larger birds for whom heat loss is slightly less critical. Species like Dunlin and Ringed

Plover are therefore more susceptible to severe winter conditions than Curlew, or Brent Geese and swans.

As winter conditions take a grip on the mudflats, food supplies become more critical than at any other time, but food is more difficult to find. Daylight hours, neccessary for species like Redshank to hunt for food, become short, and when high water comes in the middle of the day, the period when shoreline food is both visible and uncovered may be very short. Thus for Strangford's birds, twice a month there is a period of several days when it is extremely difficult to find enough food, although some birds like the plovers can feed at night. Even when the tide is out, finding food is not easy. Severe weather may force the birds into areas which provide some shelter but are less productive for feeding. Some prey species, Lugworms, Ragworms, and the Baltic Tellin burrow deeper in winter, particularly during frosty conditions, and so they become more difficult to find. Furthermore, as winter progresses the constant foraging of the birds reduces their food stocks.

In the very severest weather, even Strangford Lough with its mild

A Redshank hunting for small worms and shrimps before the rising tide. Numbers in the Lough have increased in recent years (L.J.T.).

Black-tailed Godwits (Limosa limosa) *foraging in salt marsh (L.J.T.)*.

climate, can suddenly become inhospitable when the shore ices over. Such conditions can occur when a low tide at night allows a thin film of water on the shores to freeze; this stays frozen during the high tide, and a new layer freezes again during the next low tide, so gradually accumulating a bed of rather soft, crunchy ice up to a foot thick. The northern mudflats near Ballyrickard seem particularly prone to this, but in recent years it has occurred in many areas, most notably near Castle Espie, Ardmillan, the approaches to the Quoile Estuary and in Castle Ward Bay, all areas where fresh waters from local streams and rivers reduce the salinity.

When freezing occurs, the risk of severe mortality to the birds is considerable, because inland areas, and neighbouring coasts may be in a similar state, so alternative sites are not readily available. Perhaps the most obvious signs of stress can be seen when birds accumulate along any part of the shore where ice has not collected. Land birds like Redwings, Fieldfares, Starlings and even tits, finches, and buntings may feed alongside Curlew, Dunlin, Knot and Redshank as well as a mass of Black-headed Gulls (*Larus ridibundus*), all jostling in the restricted space for a harvest of partially frozen worms and shrimps buried in the banks of storm tossed weed. It can be a time of plenty for the more spectacular predators; Peregrines (*Falco peregrinus*) and Sparrowhawks (*Accipiter nisus*) maraud the flocks, singling out individuals straying away from these great concentrations. Foxes, Stoats, and even Otters may visit the icy shore in these conditions, stalking the weaker birds, or scavenging those already dead.

What to Eat

One might be forgiven for supposing that the large variety of birds arriving to overwinter in the Lough would cause excessive competition for the available food. In fact, they manage to co-exist in some degree of harmony. This is because each species has its own particular feeding requirements, met either by choosing particular food types, or selecting items within a certain size range, or by feeding in different areas. As a result, problems of direct competition between species are kept to a minimum.

The nature of a species' feeding habits can often be seen in the shape of its bill. Wildfowl have rather broad, spade shaped bills equipped with a series of plates capable of sieving food from water. They tend to be dabblers or grazers, but within this group, comprising the swans, geese and ducks, there is considerable variation in the types of food taken. Brent and Wigeon spend part of their winter grazing on the eel-grass swards that cover large parts of the northern mudflats in early Autumn. The effect of these flocks on the area is considerable, and by the end of November most of the main stretches have been stripped of their eel-grass, the birds switching to the green alga *Enteromorpha*.

Whooper and Mute (*Cygnus olor*) Swans dabble for grass fragments, seeds, fine roots, eel-grass and small invertebrates. To obtain this food they, like the Brent, often indulge in an activity that is worth looking out for: puddling. The best way of seeing how this works is to go down to the shore, for example at Mount Stewart where the sand is firm but fairly moist, and tread up and down on the same spot for several minutes. Soon that spot will become very soft, almost slushy, with a debris of small bits of vegetation and seed slopping around on the surface. The birds do precisely the same thing, and then go one stage further and scoop out the food. In late Autumn and Winter, when food is beginning to be short, it is possible to see signs of this activity on many shores in the north of the Lough.

Places where farmland grain has been washed onto the shore are particularly attractive to wildfowl. In early Autumn after a good harvest, these areas are easily found, many associated with fresh water streams which themselves are an added bonus for the birds. Ballyreagh lay-by, near the Newtownards Maltings, is a classic spot for this; the view north will reveal numbers of Mallard, Pintail, Brent and Swans all dabbling for the barley washed out in the Maltings Stream. Other such places can be seen at Ardmillan and Ringdufferin (both good for Teal in frosty weather), Castle Ward Bay, and on the eastern shore of the Lough at Salt Water Bridge near Kircubbin.

As the rising tide gradually covers their feeding grounds many wildfowl can be seen "upending" to get within reach of their food. For the Brent it may often be easier for them to feed this way when the eel-grass is wafting back and forth in the water, rather than having to pick at it when it is lying flaccid on the exposed mud.

Amongst the wildfowl there are some specialists. The brightly coloured Shelduck (in the sun, often the most conspicuous of the wildfowl) prefers to dabble for the Laver-spire shell. The Shoveler (*Anas clypeata*) with its enormous bill tends not to upend or graze but prefers to paddle through shallow muddy water, bill thrust forward, collecting small bits of food debris and sieving out the water. Two diving ducks, Goldeneye (*Bucephala clangula*) and Tufted Duck (*Aythya fuligula*) occur widely in the Lough, particularly in the north. These specialise in diving for invertebrates in the waters off the mudflats. Pochard have a similar lifestyle, but normally are found in the fresh water areas like Greyabbey lake and the Quoile Pondage.

A number of wildfowl species particularly favour farmland near the Lough. The large open fields of fine quality farmland close to Comber and Greyabbey seem especially attractive to them, though they can come

Whooper Swans. Mallard and Wigeon dabbling for grain washed out from the Maltings at Ballyreagh, Newtownards.

Brent Geese in Castleward Bay, near the Black Causeway. This area is popular with them for its stocks of Enteromorpha *in later winter.*

on to land almost anywhere round the Lough, or on the larger islands. Feral Greylag and Canada Geese (*Branta canadensis*) are often seen in these areas, with very occasional Greenland White-fronted Geese (*Anser albifrons*), whilst Mallard and Teal home in on any areas of stubble flooded after Autumn storms. Particularly notable is the recent increase in Whooper Swans using the large fields of barley stubble between Comber and Scrabo Hill. The number of fields given to barley production appears to have increased in recent years and this, together with the destruction of many of the field boundaries, has made the area very attractive to these birds which commute back and forth from roosting on the Lough to exploit the quantities of spilt grain left behind by the combine harvesters. It must be one of the very few examples of wildlife actually benefitting from intensive modern methods of agriculture!

Waders feed on a wide variety of small worms, crustaceans and molluscs. Perhaps more than any other group of birds on the Lough they illustrate the ways in which the basic 'idea' of the bill has been adapted to different methods of feeding. The long down-curved bill of the Curlew for example, is particularly suitable for withdrawing Lugworms from their deep burrows, whilst the robust chisel-shaped bill of the Oystercatchers (if you find a dead Oystercatcher, just feel that skull!) is ideal for breaking open Cockle and mussel shells. Knot and Dunlin by contrast, have rather short light bills suitable for picking food from the mud

The Wintering Birds of the Shore

surface or just below it; small Corophid shrimps and very young bivalves form a major part of their food. Birds with long narrow bills like the Godwits would be incapable of turning over stones to find food, but the Turnstone, as its name implies, has a broad, very slightly upturned bill that is ideal for this purpose.

The situation is however, rather more complicated than this. Why should the Curlew's bill be down-curved, whilst the Bar-tailed Godwit, feeding on much the same food species, manages to get by quite well with a slightly up-turned bill? This is still a mystery, but an interesting suggestion for the Curlew has been made. The idea is that the curve of its long bill allows the bird not only to penetrate the sand burrows of the Lugworms, but to search a greater area in the sand. The Curlew does this by pivoting round the point of entry with the bill still buried in the sand. Whether this actually is a real advantage remains to be proved, but that type of feeding activity certainly does occur. Examination of Curlew probe holes in firm sand will reveal a ring of footprints round the hole as the bird works its way round, and indeed anyone watching the mudflats dotted with Lugworm casts will soon spot the Curlews, their bills buried full length, shifting from side to side 'sweeping' for the worms below.

Perhaps one of the most elaborate feeding methods to be seen around the Lough is employed by the Oystercatchers. The process starts with the

Roosting Curlews; one of the main predators of Lugworms (L.J.T.).

Oystercatcher following the retreating water's edge on a falling tide. While the sand is still waterlogged and soft the bird walks along slowly, bill buried in the sand, constantly probing up and down for a Cockle. It does not take long to find one in a good area like Greyabbey or Kircubbin Bay. With a sideways flick of the head the bird levers the Cockle to the surface, and then picks it up and carries it to an anvil not more than a few metres away. These anvils are easily found - scatters of empty shells round stones or patches of hard sand - on occasions they can even be spotted from the warmth of a car in Greyabbey car park. At the anvil the Oystercatcher rapidly and repeatedly hammers the Cockle, and after a few moments the shell is cracked open, and with a twist of the bill the meat is consumed. Finally, the bird goes off in search of another Cockle. In this way astonishing numbers can be consumed. One bird can take over eighty Cockles in an hour, and with usually over a hundred Oystercatchers feeding in the Bay in the winter, over forty thousand cockles are taken in a single low tide at Greyabbey alone.

Other areas also provide a rich winter harvest. The banks of dead seaweed on the strandlines constitute a rich hunting ground at all stages of the tide, so they are particularly useful. The ubiquitous Redshank, present in internationally important numbers, is something of a specialist in this area, feeding on amphipods and other small invertebrates. These provide a major supply of food, not only for the Redshank, but also for the Turnstone, often seen flicking over the dead weed, and remarkably adept at balancing and running over strandlines floating in the waves of a spring tide. By contrast, Black-tailed Godwits (*Limosa limosa*) specialize in areas of very soft mud where they take a large variety of invertebrates and even plant material.

Other good areas for waders include the rocky shores at the southern end of the Lough. Here again, the Turnstone makes its appearance in large numbers, often along with Purple Sandpipers (*Calidris maritima*) and Ringed Plovers. Sheltered shores with brackish streams and pools are often used by Greenshank (*Tringa nebularia*), usually in ones or twos, and they are one of the few waders to be seen taking small fish. Land adjacent to the Lough is also extremely important for waders, not only for roosting during very high tides but also for feeding. Curlew, Lapwing (*Vanellus vanellus*) and Golden Plover may all be seen feeding for worms and small insects on pastureland near the lough, whilst the adaptable Oystercatchers can often be seen taking earthworms on football pitches within a few minutes' walk of Newtownards town centre.

In addition to wildfowl and waders, other types of bird make regular use of the Lough shores, and in some cases their activities are not confined solely to the winter. When the tide is out, large flocks of

The Wintering Birds of the Shore

Signs of good hunting; footprints left by waders in the Comber Estuary.

Starlings come to forage, and although there is little information available, their numbers are sufficiently large to merit their inclusion in any list of major shoreline feeders. Rock Pipits, although not nearly so numerous, are one of the most widespread species, hunting for small crustaceans out on the rocks, in banks of dead weed, and on the shingle and salt marshes. Hooded Crows (*Corvus corone corone*) and the much scarcer Carrion Crows (*C. c. cornix*), as well as Rooks (*Corvus frugilegus*) and Jackdaws all resort to feeding on the shore. They have adapted the Herring Gull technique of carrying shellfish high into the air and dropping them onto stones to break open their shells. Herons, with their spear-like bills are a common sight in the south of the lough. Poised motionless by the water, they wait for small fish to appear. When the tide is in, Kingfishers are regular hunters of small fish and shrimps in sheltered corners, and are quite widespread about the Lough. Such areas are sometimes frequented by the Little Grebe, or Dabchick, diving and hunting amongst the floating weeds.

Getting Enough Food

Whatever the type of bird, and whatever the type of food and feeding technique it employs, the dominant requirement in winter is to eat as much as possible to replace the energy spent in keeping warm. The Lough's wildfowl and waders, with their dependence on the tides show this particularly clearly.

Drumlin islands off Whiterock provide rich winter grazing for wildfowl.

As the shoreline gradually becomes exposed by the ebbing tide, flocks of birds leave their roosts and make their way towards the water's edge. Brent Geese, roosting afloat on the high tide, start upending as the falling water level brings them within reach of the eel-grass. Then, as the mudflats become exposed, they come ashore, grazing in flocks of several hundred or more. Waders come in from roosts, or adjacent fields and saltmarshes, settling on the shore and gradually dispersing as greater areas of mud are exposed. As the mudflats dry out, other species - Mallard, Teal, as well as Mute and Whooper Swans, converge on streams flowing over the shore or around pools retained by banks of sand.

In the time allowed by the ebbing tide all these different birds have to graze, dabble, or catch enough prey not only to support them over the next tide, but also possibly through a bleak winter's night, or even severe storms. As we have already seen, it becomes progressively more difficult for most species to find food as winter advances. How do the birds make best use of the time available to them when the tide is out?

Any animal faced with a declining food supply has several options. It can go and feed somewhere else, and for the birds in Strangford Lough this might be achieved by shifting from one bay to another, or even leaving the Lough altogether. That would help reduce the amount of competition for food for those left behind. Alternatively an animal might alter its feeding habits. For the birds on the Lough this might entail changing to a different food type, or 'fine tuning' their existing food selection to get the best of what is available, for the least effort. The best way to see how this works is to look at two of Strangford Lough's wintering species, the Brent Goose and the Oystercatcher, and see how they make the most of their food stocks.

The Brent Geese arriving in early Autumn find swards of eel-grass that have been growing on the mudflats since early summer. Although patchy in places, this green pasture extends from Island Reagh, past Castle Espie and Comber up to Newtownards and then southwards to Greyabbey. There are also other areas in Mahee Bay, Ardmillan Bay and on the eastern shore, for example around Gransha Point. By October the growth of the eel-grass has ceased, but at this time the food stocks are rich, and the geese concentrate on the most extensive swards in the north of the Lough. As their numbers build up with fresh arrivals from across the Atlantic, the demand for eel-grass rapidly increases, particularly as by now Wigeon are also grazing in some of these areas.

In October (formerly, November) numbers of Brent are usually reaching their peak with as many as fifteen thousand geese feeding alongside Wigeon that are still arriving from Iceland and later, eastern Europe. By this time the stocks of eel-grass are declining rapidly; the colour of the mudflats from the distance is gradually resuming its grey-brown tone. The very large flocks begin to break up, some remaining in the north of the Lough, others increasingly making use of Greyabbey Bay, Chapel Island and Mount Stewart, whilst on the western shore they move into Mahee and Ardmillan Bays. In spite of the need to move around, the Brent still remain in family parties, the white bars on the wings of the young still easily visible. By this stage however, some groups find the effort in searching out eel-grass has reached the point where the reward is no longer adequate, and groups begin to leave the Lough. They may make an appearance in Dundrum Bay or Carlingford Lough, but the most likely venue is Dublin Bay where they accumulate in large numbers some time before Christmas. Others may continue on to Wexford or Cork. If recent breeding in the arctic has been successful, and larger numbers are present, the departure from the Lough may be more rapid as the food is consumed faster, whilst lower numbers in any particular year may allow the flocks to linger on for a few weeks more, by virtue of the continuing food supply. As a result, the actual amount of eel-grass consumed each winter may not vary as much as changes in bird numbers might suggest.

For the Brent remaining on the Lough, the flocks break up still further, appearing at Gransha Point, Castle Ward Bay, and many other places in the south of the Lough. At this stage, not only have many birds left the Lough, but another change is taking place. Gradually the Brent abandon their prefered eel-grass, and start consuming *Enteromorpha*, and by the New Year this forms a major part of their diet. Of all their food plants it is much the most widespread, and it is quite possible that in spite of the birds' preference for eel-grass, *Enteromorpha* is more important as a food source throughout the whole winter.

Finally, a further change in diet may occur in the latter part of the winter. Small groups may be seen feeding on saltmarsh plants immediately above shore or even moving onto some of the larger islands to graze on land grasses. This is particularly the case on some of the islands off Whiterock, with their wide open pastures reaching down to the shoreline. This trend has increased in recent years to the point that on some occasions several hundred geese may be seen using the islands in this way. A similar situation occurs in Dublin, when the geese move off the mudflats to exploit the lush grazing to be found in the parks and golf courses, within a few yards of the city traffic!

Oystercatchers employ a similar strategy but there are important differences. Examination of a collection of dead shells from any of the 'anvils' in Greyabbey Bay or off Rough Island in autumn reveals that the shells are remarkably uniform in size; about 2cm long. Yet on the shore there are living Cockles ranging from the size of a very small pea through to individuals about the size of a plum. Clearly there is a selection procedure involved here. The birds are ignoring the tiny cockles probably because there is not enough meat to make it worth opening the shells, whilst the old, very large meaty cockles are so rare that unless they are found by chance, it is not worth hunting for them. Middle-sized cockles provide a compromise, and we can see that the hunting grounds, clearly delineated by the piles of dead shells, reflect the best spots to go hunting for them.

Cockles do not grow much in the winter; indeed, they can experience their own food shortages as the amount of dissolved nutrients and plankton decline, and they may lose weight. Thus whatever the stock of Cockles and their meat is in the autumn, it has to last until the following spring. As with the Brent therefore, the amount of food available to the Oystercatchers gradually dwindles over the winter.

The Oystercatchers initially respond to this problem by congregating in the more productive parts of the Lough, as the poorer areas become depleted first. Secondly they identify which parts of the shore are best (possibly by discerning the distribution of opened shells) and concentrate on these. The end result is that the birds concentrate more and more tightly on these areas of suitable Cockles, and as the Cockles get eaten each area gets smaller. All this helps to maintain the food intake for a while, but sooner or later (as with the Brent, depending on the number of birds feeding) supplies must run out, or at least become uneconomic in energy terms. When that happens, two changes take place. Firstly Oystercatchers may switch to Lugworms or Baltic Tellins. This is often when they start hunting on adjacent fields and football pitches - obviously not to find shellfish, but because the supply of earthworms is now more rewarding. Secondly, the size of the freshly killed Cockles left around the anvils begins to change. The Oystercatchers, under the 'beggars can't be choosers' principle, relax their size criteria and generally accept smaller Cockles. Having made this adjustment, new areas now qualify as hunting grounds and we can see a corresponding expansion of feeding activities to much wider areas of shore. The Oystercatchers may even start selecting Shore Mussels (much less rewarding than cockles in meat content) if the weather is severe. This can be confirmed by looking for a scatter of empty mussel shells round the low water rocks in places like Greyabbey.

So, to summarize, although the Brent and the Oystercatchers have very different foods and methods of feeding, they employ similar strategies to get the most out of Strangford Lough in winter. Upon the success of these tactics depends not only their fitness to migrate north and breed at the end of the winter, but also their very survival.

A Good Rest

With the turn of the tide, the water's edge starts to advance once more. On the steeper, boulder strewn shores this may take the form of a gradual progression, but on the flatter, muddy or sandy shores the tide advances in a series of fits and starts, depending on almost imperceptible undulations of the sediments. On level areas the water may advance very rapidly, gaining a hundred metres or more within a few minutes.

The birds work their way along the water's edge, gradually retreating as they try to get the most out of the feeding available to them. Family parties of Brent Geese almost seem to leap-frog each other to get at new areas of food, leaving the others closer to the water's edge. As the flock shuffles backwards before the advancing tide, minor disputes over feeding patches become common. If one member of a family comes too close to the 'feeding zone' of another, one of the parents may charge forward a few yards, head low, neck stooping almost parallel to the mud, and as a rule the intruder retreats.

Knot, Dunlin and Ringed Plover run along the water's edge dodging back and forth between waves, the former justly earning its scientific name *canutus*, the whole flock hunting and picking for small worms, bivalves and shrimps that are moving closer to the surface as the shore becomes waterlogged once more. After resting for a while during low water, Oystercatchers resume their hunting, finding the sand easier to probe as it becomes moist.

Eventually flocks of birds start to move around the Lough as the momentum of tidal advance begins to pick up. Wildfowl adapt regular flightlines, adjusting these to the timing of the tide and to the weather. Brent in particular are a regular sight flying northwards from Greyabbey and Mount Stewart, these areas becoming covered in advance of the most northern mudflats. On the way they join up with flocks from the Boretree Islands and Cunningburn. Finally, from the lay-by at Ballyreagh, or the tide bank near Newtownards airfield, particularly in early autumn, it is possible to see thousands of geese converge along the northern shore as far as Rough Island, and mix with flocks arriving from other parts of the Lough. Some wildfowl like Mallard and Teal may leave to visit ponds and wet hollows inland, whilst birds in the south of the Lough may move into the Quoile Pondage.

Eventually the most northern corners of the Lough are submerged. By this stage there may be over ten thousand birds floating offshore. Brent form the bulk of these, but amongst them may be Wigeon, Mallard and Pintail. Similar flocks of Brent are also be seen in Greyabbey Bay, the Comber River and off Castle Espie. The noise at such times is spectacular - perhaps best heard whilst stumbling along the Newtownards tide bank

Dunlin roosting on salt marsh at Gull Rock, Mahee Island (L.J.T).

at night! The constant honking of the Brent, unseen across the dark water, pierced by the shrill whistles of the drake Wigeon is an eerie sound, especially coming as it does, from so close to a town centre.

The response of the waders to the rising tide is no less spectacular. Some species like Bar-tailed Godwit may remain close to the water throughout the tidal cycle, and as the tide rises precede the advancing water's edge with a series of brief runs whilst continuing to feed. This probably reflects the widespread nature of their prey, since they feed on a range of different species distributed over much of the shore. Others with more patchy prey distributions may remain in their preferred feeding areas until the sand is completely covered. When this happens the birds prepare for roosting, congregating on shingle banks, small islands, areas of salt marsh and fields near the shore. Turnstones and Redshank continue to forage along the strandlines, but the majority of birds will rest, sleeping and digesting prey taken a few hours before, and occasionally pecking for food in a desultory fashion. Some species, Oystercatchers for example, may resort to Belfast Lough which is only a short flight away. Here the tidal cycle is out of sequence with Strangford, thus affording opportunities for further feeding.

Before arriving at their roosting areas, some waders form sub-roosts out on the sand. In some estuarine areas round Britain and Ireland these are apparently well defined, composed of densely packed accumulations of one or two species. The situation in Strangford Lough does not seem to be quite as clear cut, and sub-roosts often appear as rather loosely-knit assemblages of birds which often continue to pick at food on the mud. Golden Plover and Knot are exceptions however, and their tightly packed flocks can often be seen clearly on the mudflats in the vicinity of the Butterlump Stone south of Newtownards.

About an hour before high tide, it is time for bird watchers to get in position to await the arrival of the birds at their roosts. Because of the concentrations of birds packed into small localities, roosting is perhaps the best time from the observer's point of view, though of course the opportunities for watching the birds actually feed are restricted. It is important to have a tide table. Apart from giving the times of high water (allow for about 30 minutes difference between Killyleagh and the north of the Lough) the height of the tide is important, as a neap tide may allow the birds to continue feeding out on the mud far away from the roost, whilst a spring tide may leave the observer soaking wet.

Roosting sites are distributed widely about the Lough, but the extent to which a roost will be used on a particular day depends on the weather, local disturbance factors, and the height of the high tide. Many regularly used sites are quite easily observed from land and make interesting bird watching. Greyabbey Bay for example, is an excellent location where a low bank of shingle opposite one of the car parks is used as a roost by Bar-tailed Godwit, Oystercatcher, Golden Plover and Brent Geese, as well as other species, in all but the very highest of tides. The saltmarsh at Castle Espie is another good area, whilst the outcrop of boulders in the middle of Mahee Bay is regularly used by Black-tailed Godwits, Shoveler and Pintail. In the extreme south of the Lough Mullog and Killard Points are both regularly used as roosts.

There are also a whole series of small saltmarsh areas, islands and small pladdies where waders may roost; in the Boretree Islands alone there are over ten such sites where Grey Plover, Turnstone, Knot, Dunlin and Oystercatcher may come, the latter often in their thousands. Swan Island in Strangford harbour is an excellent site for seeing roosting Lapwing, whilst the small saltmarsh island north of Whiterock is good for viewing Teal, Brent, Bar-tailed Godwit and Snipe. In fields immediately inland from the shores, Curlew, Golden Plover and Lapwing may continue feeding throughout the high tide period, and at very high tides they may be joined by Bar-tailed Godwit and Knot.

As the tide creeps up over the subroosts and the last patches of feeding ground, the flight of the waders to their roosts is spectacular. Flocks of Knot numbering several thousand birds rise and twist through the air, by their very presence attracting other birds to join the display. Golden Plover and Lapwing circle high over small inlets and their adjacent farmlands - look for these high over the Comber River. They are specialists in land feeding, and their soaring flight is often the preparation for an incursion into the nearby open fields. Oystercatchers, Bar-tailed Godwit and Curlew arrive at the roost in more ragged groups, often patrolling up and down an area before finally settling. Small groups of Dunlin, often with Ringed Plover and Turnstone, fly back and forth low over the water's edge, before choosing their roost, turning into the wind and quickly settling, their wittering calls finally ceasing as they get comfortable.

One by one the different flocks accumulate, larger waders like the Curlew standing head and shoulders over the densely packed small waders. Some continue to feed briefly, Grey Plover pecking for suitable items between the stones, dead weed and saltmarsh plants. Soon however, the flocks settle down; many birds sleep, bill tucked into feathers, whilst others preen or merely digest their food, finally regurgitating crop pellets composed of the indigestible portions of their prey. All birds face into the breeze, many with one leg tucked into the feathers.

During spring tides many of the roosts may be covered and inaccessible to the birds. Under such conditions some flocks may stay aloft for a major part of the high tide period, circling high over the bays and salt marshes, in effect, 'roosting on the wing.' Alternatively they may settle upon fields near the lough shores, particularly if recent ploughing holds out the promise of extra feeding. Such flocks may often be seen near the Comber - Newtownards road where it passes near the tide bank.

Why roost at all? A number of explanations have been suggested, and they may all be valid. Firstly, predators tend to be confused by very large flocks or groups, of prey, so for small birds (or even shoals of fish and

---

CONSERVATION OF WINTER BIRDS

Conservation of winter birds formed the initial impetus for the establishment of the Strangford Wildlife Scheme. This involves the control of most of the shore of the Lough by the National Trust either through ownership or leases (see Chapter six). Shooting, previously uncontrolled, is regulated by a system of permits, and areas in which shooting is prohibited by agreement with local wildfowlers. The primary aim is to provide extensive areas in which overwintering wildfowl and waders may feed, roost, and obtain shelter. So far as it is possible the Trust seeks to control all types of disturbance in these areas, whilst managing the others in the interests of wildlife conservation, but accepting a certain amount of compatible human activity within them. Bird numbers have been carefully monitored since 1965 to check on the success of these policies.

Increased attention is now also being given to the managment of marginal areas round the Lough and on the islands for bird life. This may involve the managment of grazing on the islands to produce good grass swards for wildfowl, or the creation of wet areas on land for the feeding and shelter of both wildfowl and waders.

---

swarms of insects) there is safety in numbers. This seems to work for some waders around the lough, for most of those killed by Peregrines for example, seem to have been caught whilst feeding on the mud flats, rather than during roosting. If this is the case, the enormous and highly conspicuous flocks of waders are actually safer places for them, and they can select strategic locations from which to resume their feeding during the next ebb tide.

Safety in numbers does not always work, however. Short-eared owl pellets collected from a frequently used roost in the north of the Lough have twice been found to contain the rings used to record bird movements; astonishingly these were both from Dunlin ringed by the North Down Ringing Group almost seven years previously, and bearing adjacent numbers in the sequence. This suggests both regular predation by the owls, and a strong cohesion in the flocks over a long period.

Secondly, the fact that individuals often change from one flock to another as they accumulate and mingle in a roost, may serve to even out the food supply. A family of Brent, for example, having had a somewhat lean time foraging for the last eel-grass off Rough island may well 'decide' to join a different flock that is apparently faring better, and end up going to Greyabbey instead where the feeding might still be quite good. Precisely how the mechanism actually works is not clear, but some evidence for this interchange comes from Brent marked with coloured rings that have changed between flocks within the Lough and even further afield. If this occurs regularly, it would certainly reduce the risk of any one group of either wildfowl or waders doing badly.

The birds also need time to rest and digest their food. This is important, particularly in the case of the waders. They need to regurgitate pellets containing the more indigestible portions of their prey, which may be shell fragments, grit, or the mouthparts of some of the larger worms. High tide is the best time to do this, since the feeding grounds are covered anyway, and it is not energetically worthwhile for most species to move elswhere during the high tide.

Thus although roosting appears to be a period in which the birds are doing very little, it is in fact extremely important in their cycle of activities, and as many as ten thousand birds may be congregated in a single site. It is therefore important, particularly during severe weather, that disturbance of roosting sites be avoided as much as possible.

The Return Journey

The increase in daylight hours and the gradual improvement in the weather finally herald the onset of spring.

By contrast with their arrival, the departure of the wintering birds is an almost discreet affair. There may be a logical reason for this. With the reduction in food stocks, flocks of many species may be reduced in size as they break up to forage in smaller, less productive locations. As the birds leave the Lough, they are replaced by other small parties coming in on their way north, and the flow of birds gradually declines to a trickle, and finally ceases. The very large flocks of autumn with their bulk demand for food simply could not stay together in this time of heavily depleted food resources, although on occasions there are large influxes of passage migrants stopping off for a brief refuelling visit.

Sometimes birds can be seen actually leaving. Brent Geese, unlike their autumn arrival high in the sky over the north of the Lough, may be seen flying out at low level past the Angus Rock and Ballyquintin Point, presumably to swing north and make their way along the coast before the final break for Iceland and the arctic beyond. Other species of wildfowl

One leg tucked in to keep warm, a Ringed Plover awaits the turn of the tide (L.J.T.).

like Teal, Wigeon, Shoveler and Pintail, also desert the Lough, along with most of the waders. At the same time, in keeping with the general northward shift of the birds, the wintering Oystercatchers and Ringed Plovers moving north are replaced by individuals from populations further south, continuing the presence of the species into the spring and summer. Individuals of some species hardly move at all, so that Shelduck, Mallard and Red-breasted Mergansers may be considered year round residents of the Lough, although their numbers are boosted in the winter.

So, finally the winter activity on the mudflats dwindles away. In a sense it is a sad time, as these areas become quiet, but it is an essential part of the annual cycle. The winter has been a frenzy of constant foraging and probing by the thousands of waders and wildfowl. Now, the warmth of the coming months, and a period of relative calm, is essential to allow the shore life to recover, grow, breed and flourish. After a few months the first hesitant blades of eel-grass begin to poke through, a winter ring on Cockle shells marks the edge where new growth begins, and many species fatten and develop in preparation for spawning. Only in this way can the Lough's shores be ready to support the wildfowl and waders in the next winter.

8 *Nesting Birds of the Lough*

THE START of the nesting season, with all its promise of future generations, could hardly be less auspicious. Winter is entering its bleakest phase, with food stocks dwindling, and harsh conditions demanding the most from the birds. Parties of Brent Geese are still scattered over the mudflats and in the small bays and channels, whilst large flocks of waders still circle high over the shores before the advancing tides. The flocks of birds are relying on their collective abilities to exploit the food resources as effectively as possible. But in the midst of all these challenges to survival, a subtle change is beginning to take place, in which very different strategies come to the fore.

This is a time for establishing new relationships or reaffirming old pair bonds. Some birds, like Oystercatchers and other waders in the large flocks begin to appear as pairs. Male and female Mallard fly together amongst the ponds and small bays around the more sheltered islands. Red-breasted Mergansers begin to hunt and dive in pairs amongst the floating weeds. High in Scots Pines, Herons begin the laborious task of repairing their massive nests, carrying almost impossibly large sticks and branches in their bills.

Mild spells during the winter give added impetus to the process. Two or three weeks of frost-free weather is sufficient to coax Herring Gulls to to start desultory nest scrapes with encircled grasses and bits of weed in the more sheltered corners of the islands; but in spite of these early starts, their first eggs will not be laid until late April. However, mild winters can result in true early nesting for some species; a Mallard nest with over ten eggs was found near Strangford in mid-January 1989, along with a scatter of reports about early nesting garden birds. Such early nesting is of dubious benefit. Even if severe conditions do not reassert themselves, the general cold and damp threaten the success of early eggs, and once hatched, there is often a lack of food for the young birds.

Many wintering migrants indulge in early courtship and displays in preparation for nesting in their northern breeding grounds. Goldeneye, in spite of the winter squalls, begin their courtship, the males competing for the attentions of a female, their black and white heads bobbing up and down as they circle round in the cold choppy water. Brent Geese,

Lythe Rock near Mahee Island. Blackthorn scrub provides nesting cover for wildfowl, whilst open grass and shingle support gulls and terns.

often faithfully paired for long periods of their lives rejoin their partners after temporary separations foraging for food in different parts of the Lough, or indeed, of Ireland. They mate before setting off for the Canadian Arctic. The previous year's juveniles begin to exercise greater independence from their family - although in some cases tenacious youngsters can hinder the efforts at further nesting by their parents.

The Lough as a Place for Nesting

Gradually the Lough begins to emerge from winter. In spite of the false starts and setbacks caused by the unpredictable weather, by mid-March many of the Lough's year-round residents have begun nesting in earnest. In doing so, they begin a slow process whereby the whole emphasis of the bird life shifts away from winter feeding and flocking along the shores, and moves out to the islands where the main nesting effort will take place. The large number of islands, of many shapes, sizes and types, results in the Lough being a particularly important resource for many species of nesting birds. For some species like the terns, the area is of international importance.

There are several reasons for the importance of these islands. One is security from disturbance and predation. Only the largest islands like

Taggart or Conly, can support Foxes or Badgers. Generally, the smaller the island, the less likely it will have major predators (except where they can swim or walk across at low tide), although there are few islands that are completely free of Brown Rats. No islands of course, are immune to airborne predators like gulls, Hooded Crows or Peregrines, but as we shall see later, many nesting birds derive considerable protection from camouflage, concealment, and safety in large numbers. The isolation of the islands from the mainland also provides some freedom from human disturbance, particularly in the spring when there are few boating people about on the Lough.

Another reason for the importance of the islands for nesting birds is that they provide a variety of essential habitats. Dense covers of Alexanders, Nettles, thistles, and rank overgrown grasses are ideal for nesting wildfowl. Often there is good grazing near the nest site; feral Barnacle Geese (*Branta leucopsis*) for example, trim the grass in patches near their nest. Ponds provide the young birds with water, and in early springs the relatively mild conditions help to develop an early supply of insect life. Elsewhere, the fringes of Common Salt-marsh Grass, Scutch grass, Scurvey Grass and Sea Mayweed, and the stretches of stones and shingle are ideal for terns and waders, whilst gulls seem to be able to exploit almost any island habitat, nesting in grass, amongst boulders, on stone walls and in dense weeds.

The islands are also essential as bases from which to depart for hunting and feeding. This is not merely a matter of simple food aquisition, because obtaining food for hungry chicks requires an investment of energy, and clearly the less effort expended in a hunting or foraging expedition, the greater the 'profit' obtained from the food acquired. In some parts of Britain the initial investment can be enormous. In Strangford Lough however, the rich plankton content of the water supports very large populations of Sand Eels and Herring fry close to many of the island colonies, thereby reducing the time and energy spent travelling to a minimum; a considerable benefit for parents faced with the urgent demands of hungry chicks. Equally, Cormorants nesting in the Lough benefit from fish populations adjacent to their colonies.

So this is the resource for the nesting birds. A wide variety of islands, habitats, and locations, many of them in close proximity to rich feeding areas. Up to forty different species of bird regularly nest on these islands or near the mainland shores of the Lough. In doing so, they employ a number of different strategies to achieve successful nesting, involving differences in their choice of nest site and construction, feeding, and the development of their young.

Wildfowl: a Strategy of Concealment

Wildfowl as a group have the longest nesting period of all the commoner breeding birds on the Lough, extending from the cold of late winter well into the warmer weather with relatively late nesters like Shelduck and Red-breasted Mergansers. The first species to begin nesting in earnest are usually the Mallard. By mid-March preparation of their tightly woven nests is well advanced. Mallard nests are difficult to spot, as they are usually constructed in thickets of Brambles, Alexanders or long grass. Often the only indication is a small track of trampled vegetation leading from more open ground into the thicket, marking the regular route used by the birds. If disturbed, the parents scuttle through the vegetation a short way before breaking into flight, thus concealing the precise location of the nest. Up to fifteen creamy-white eggs are laid, although one nest has been found with twenty-four eggs - probably shared accommodation! The eggs are kept concealed and maintained at a regular temperature by quantities of down plucked from the female's breast.

Suitable ground with sufficiently dense cover is normally only available on the larger islands, like the Boretrees, some of the islands off Whiterock, and places like Jackdaw and Dunnyneill Islands in the south of the Lough. Up to fifty nests are found on the islands each year, and the trend seems to be one of steady increase, but this figure is probably an underestimate, since a really thorough survey would cause too much disruption. On mainland areas, almost any damp ditch or bank of weeds may support a Mallard nest, though Brown Rats, Foxes, and humans with their pets are a strong deterrent.

Mallard are the most numerous duck nesting on the Lough, although there are a number of other important duck species, that nest somewhat later in the season. Red-breasted Mergansers nest in the dense fringe of vegetation round islands like Jackdaw and Dunnyneill, laying about twelve pure white eggs in April. Shelduck nest widely around the Lough both on the mainland and on the islands, often in disused rabbit burrows and other hollows in the ground. Other species nest less frequently. Tufted duck occasionally build small nests tucked into the fringe of Alexanders thickets. Eider (*Somateria mollissima*), formerly a very scarce species on the Lough, but now seen quite frequently in winter, established a nesting foothold on Gabbock Island in 1985, and have since nested on the Boretrees. Usually it is the females that are seen, since the males leave their partners shortly after the nest is complete, probably to reduce the demand for food in the area. It is likely they have spread from the North Down coast and particularly the Copeland Islands where up to eighty pairs nest each year. This rather handsome duck seems likely to

NESTING BIRDS ON STRANGFORD LOUGH

The most notable nesting species on the islands and around the shores of the Lough. Numbers are based on the annual nest counts recorded since 1986. These counts give an indication of nesting *effort* rather than number of successful pairs, since not all nests, eggs, or young survive the season.

SWANS
Mute Swan	*Cygnus olor*	1-2, several on nearby lakes

GEESE
Greylag	*Anser anser*	Up to 11, larger islands
Barnacle	*Anser leucopsis*	Over 10, larger islands
Canada	*Branta canadensis*	Recently increased, over 30

DUCKS
Mallard	*Anas platyrhynchos*	40-50, heavy undergrowth
Shelduck	*Tadorna tadorna*	ca 10, hollows and old burrows
Gadwall	*Anas strepera*	1-2 each year
Tufted	*Aythya fuligula*	Occasional nesting
Red-breasted Merganser	*Mergus serrator*	Up to 5, prob undercounted
Eider	*Somateria mollissima*	1-2 each year since 1985

WADERS
Oystercatcher	*Haematopus ostralegus*	Recent increase, now 90-100
Ringed Plover	*Charadrius hiaticular*	Generally 20-30
Lapwing	*Vanellus vanellus*	A few nests in nearby fields
Redshank	*Tringa totanus*	1-2 nests in most years

GULLS
Black-headed	*Larus ridibundus*	Recent increase to ca 4000
Common	*Larus canus*	Recent increase to ca 60
Lesser Black-backed	*Larus fuscus*	Over 30
Greater Black-backed	*Larus marinus*	Slowly declining, ca 70
Herring	*Larus argentatus*	Recent decline, now about 1200

TERNS
Sandwich	*Sterna sandvicensis*	Variable; ca 1200 or more
Common	*Sterna hirundo*	Over 500
Arctic	*Sterna paradisaea*	About 200
Roseate	*Sterna dougallii*	Severe decline; now very rare

OTHERS
Heron	*Ardea cinerea*	Over 70, in ca 12 heronries
Cormorant	*Phalacrocorax carbo*	Up to 150
Black Guillemot	*Cepphus grylle*	Up to 10, using nest boxes

increase its numbers in the future. Finally, Gadwall (*Anas strepera*) which have nested on the Quoile since 1984 have nested at least twice on Jackdaw Island in recent years.

By mid-April the majority of other nesting wildfowl have either built nests and laid clutches, or will shortly be doing so. Relatively few Mute Swans nest on the Lough's islands, but where they do, the nests are conspicuous, and readily seen from a boat. South Island near Greyabbey, Black Rock off Ringdufferin, are regular nest sites, whilst the nest on Dunsey Rock near Ringhaddy, is often built so close to the high water level that the whole structure has come perilously close to floating off into deeper water! Many freshwater sites have resident swans, for example the Quoile Pondage, Clea Lakes and Greyabbey Lake (often called the 'Swan Hole') where the birds regularly nest in reeds near the Portaferry Road.

Both Canada and Greylag geese nest on the larger islands. These are feral birds; in other words they have been artificially introduced and are descended from stocks that have a strong element of domestication in their history. Usually they may be found where there is some cover, although Canada Geese (introduced into Britain in the 17th century) may nest rather more boldly out in the open, their nests often betrayed by small pieces of down floating off in the breeze and caught in thistles

A nest of young Shelduck on West Boretree Island, near Greyabbey (D.A.).

A Mute Swan nesting near the high water level on Dunsey Rock, near Ringhaddy.

and Brambles. When approached by humans the parents leave immediately, although again, Canadas are the bolder of the two species, hissing fiercely at the intruder before finally retreating to the water nearby, where they call to each other in distress. Barnacle geese were first introduced to the Lough in 1951 from the Inshkea Islands in County Mayo. These small geese, in the true wild state to be found nesting on precipitous scree slopes in north east Greenland, select island sites with dense tussocks of long grass, giving them both shelter and concealment. The performance of the dozen pairs of nesting Barnacle geese on the Lough has been poor, the majority of the eggs being infertile. It has been suggested that the various nutrients and minerals essential to successful breeding are not available in the diet of the Strangford birds; more likely however, is the possibility that inbreeding within this small isolated population has reduced their basic fertility. Whichever explanation is true, in spite of repeated nesting and the production of up to eight eggs in each clutch, the population has barely kept pace with its own mortality.

After hatching, life for all these young wildfowl is difficult. Where there are large pools on their islands they may remain for some time, taking a diet of insects, algae and seeds. Often however, it is neccessary to resort to the open waters of the Lough, and in May it is possible to find families of Canada Geese or Mallard, some way out in choppy water, the parents leading a string of small youngsters battling vigorously against difficult conditions. Stragglers become vulnerable to attacks by gulls and Herons, and clearly a very large proportion of youngsters do not survive their first few days of life outside the shell.

Herons in the Treetops

Woodlands by the Lough's shores support about seventy pairs of nesting Herons in eleven heronries. This is about a third of County Down's population (there are over 600 pairs in Northern Ireland), and their numbers are slowly increasing, possibly because of the recent mild winters improving their survival. The majority of heronries are on the western shore of the Lough, and this may well be linked to the scatter of small interdrumlin lakes a short way inland. The largest is Kinnegar Wood in Delamont near Killyleagh; with over thirty nests it is the largest heronry in the County, and the third largest in the Province. There are also substantial heronries at Ringdufferin and at Isle O' Valla near Strangford.

A successful heronry is an impressive sight. It is best viewed from a distance, not only because of the importance of avoiding disturbance, but also because attempting to look straight up through a dense tangle of branches into bright light to spot the nests in the tree tops is both back breaking and unrewarding! Viewed from a neighbouring hillside however, the nests are clearly seen, massive constructions over a metre across built of twigs and branches most frequently in the crowns of Scots Pines. At Lough Cowey near Portaferry however, the nests are in Willow scrub. In some respects their nesting arrangements seem precarious for such a large bird nesting at a time when late winter breezes buffet the tree tops. However, Herons are well adapted to this since their very broad wings give them sufficient lift to allow very slow landing and launching speeds. This enables them to alight with considerable precision on particular rocks or stones on the shore, or equally on their swaying nests.

Having spent February and early March repairing their nests, Herons lay about four sky-blue eggs in their enormous platform nests. Within about four weeks the young Herons are hatching, and arrivals and departures of the parents are accompanied by the lively chattering and squawking of hungry youngsters. The eggshells are thrown out to land at the base of the tree along with a fringe of droppings, and this debris is one of the best ways of confirming that a nest is actually in use.

By the end of June the young Herons are ready to leave their tree-top nests, a precipitous glide often ending in an undignified tumble on to the ground. Until they are fully capable of flight, they are extremely vulnerable. Occasionally the young Herons from the Isle O'Valla heronry have had to be shepherded off the nearby road out of danger from the traffic.

Cormorants nesting on Black Rock, near Ringdufferin. Small islands with dense weed growths provide rich hunting for fish close to the colony.

At this stage they are still easily distinguished from the adult birds by their greyish neck, brown bill and the lack of the long spotted neck feathers and black crest that make the adults look so handsome.

Cormorants; Turret Builders

By mid-April a new phase of nesting is underway . Cormorants, from numerous areas within the Lough and further afield, converge on their colonies on Bird Island off Kircubbin, and Black Rock off Ringdufferin. There they congregate in dense groups, the white thigh patches and the scatter of white feathers under the chins, typical of birds in breeding condition, being clearly visible. The latter feature seems to vary considerably. In some years it is hardly noticeable, whilst in others it gives the birds a most distinctive appearance.

The presence of Cormorants as a nesting species on the Lough began in 1980 with twenty-eight nests on Bird Island. In spite of destruction by vandals in 1984 (or ironically, even because of it, since the birds were forced to develop Black Rock as an alternative site) the colonies have expanded to a 1989 total of over 150 nests on the two islands. The nests are substantial, turret-shaped structures, composed mainly of dead seaweed, but also land vegetation, and various pieces of litter. In time the larger nests exceed half a metre in height, and this is just as well, since a number, particularly those on Black Rock, are built dangerously close to

Young Cormorants on Black Rock, with a freshly regurgitated Eel.

the high water level and are within risk of flooding during spring tides. Usually the nests are spaced at one or two metre intervals, and the dense nesting with the attendant droppings means that the colony is easily seen from a distance as a patch of white at one end of the island.

The majority of Cormorant pairs have their clutches fully laid by the first fortnight of May. Normally there are four or five chalky white, rather elongated eggs in each nest, which both parents incubate. When they hatch, the young can hardly be described as beautiful. Naked, blind, coloured dark grey, they have a distinctly reptilian appearance, betraying the ancient ancestry of all birds. Their helplessness in the first few days of life is in marked contrast to the young of wildfowl and waders, whose chicks can fend for themselves remarkably well within a few hours of hatching, given a degree of parental supervision. The young Cormorants soon grow however, and acquire a thick covering of dark grey down feathers.

The youngsters are voracious feeders; whilst one of the parents tends the nest (this is essential, given the risk of predation from neighbouring Herring Gulls), the other hunts. This is often amongst the dense bands of kelp and wrack that fringe both Bird Island and Black Rock, and the Lough's pladdies, but they may well travel further afield in times of shortage. Later, both parents are occupied in meeting the demands of the hungry young birds, bringing back a succession of Eels, wrasses of different types, and Father Lashers which are carried back in the crop and regurgitated to the young. This is not always a successful operation, to judge by the discarded fish left littered around the nests, giving the islands a powerful smell during warm weather.

Within a few weeks the young Cormorants are mobile, and capable of delivering sharp pecks to intruders, be they gulls or humans. They can run or stumble over their island with some speed, and when in the water they can swim and hunt efficiently. Learning to fly takes a little longer however, but eventually the young birds can be seen making their first weak flights. For the next two years they will bear the lighter brown plumage of the immature bird, before developing the glossy black-olive sheen of maturity.

Arrivals and Departures

Whilst the Cormorants are building their turret nests, some species of birds are arriving from much further afield to nest on the Lough. Long distance movements of such birds depend on many factors. Availability of food en route, the birds' own physiological condition, and the weather, all affect the timing of their journey. These factors play a big part in determining the time at which the first terns are seen, freshly arrived from their wintering grounds on the west coast of Africa. The British winter weather itself is unlikely to have much direct effect on much of their journey, since the birds will be responding to the conditions they find along the other stages of their route, affecting the various wintering flocks almost anywhere between Morrocco and South Africa. However, as the terns travel northwards towards Britain and Ireland, they increasingly come under the influence of the weather systems of the north-east Atlantic. Like many other species they may well adjust their arrivals and departures on the various stages of their journey to coincide with favourable conditions.

Because of this, the precise time that the first Sandwich Terns make their appearance in the south of the Lough varies considerably. Usually the first sighting is in the first week of April or later in that month, but in 1984 it was on the 1st March off Taggart Island. Favourable southerlies bring them turning and twisting high over the Narrows and the lowlying lands of Lecale, as they enter the Lough still emerging from its winter chill. Such early arrivals are the pioneers however, and it may be as much as four weeks before really large numbers of Sandwich Terns can be seen, whilst large numbers of Common and Arctic Terns do not really build up until well into May.

With the arrival of the terns, and also the first Swallows (*Hirundo rustica*), warblers and Cuckoos (*Cuculus canorus*), the Lough becomes host to an amazing cross section of the world's bird life. Flocks of Brent Geese, Turnstones, Bar-tailed Godwit, Teal and Knot are still present as representatives of the tundra and permafrost in both the new and old worlds. Golden Plover, Greenshank, Dunlin and Curlew will soon be leaving for moorland Britain and Ireland, as well as the subarctic tundra. Together with the newly arrived representatives from the tropics, they make a curious mixture of species. The terns find the Lough cold and short of food, since the plankton content of the water has not yet developed sufficiently to support the shoals of Sand Eels. By contrast, the Brent will be experiencing warm conditions compared with the last throes of the arctic winter that will greet them on their arrival in Bathhurst Island.

The Ubiquitous Gulls

Whilst the terns are still building up numbers out on the open waters of the Lough, the gulls are well into their nesting routine on the islands. The five species constitute the most numerous and widespread group of birds to be found nesting on the Lough; vigorous and flexible in their nesting requirements, there are few islands that do not carry some potential for nesting gulls.

The first species to start nesting are the Black-headed Gulls. In recent years some four thousand pairs have nested on the Lough, and numbers appear to be increasing. Over two thousand pairs nest on Jackdaw Island alone, but it is almost impossible to be precise about this, because of the difficulty in obtaining accurate counts in thick vegetation. Elsewhere, their nesting sites are more open, amongst boulders, or on gravel and emergent Sea Asters, Orache and Thrift. The nests vary enormously in structure, according to the locally available materials. Thus on Jackdaw Island the nests are formed by a small circlet of grass, whilst on Green Island Rock there is virtually no nest preparation at all. The eggs vary from dark speckled olive brown to almost pure pale blue, a range of colours that occurs in the other gull species.

Black-headed Gull colonies are noisy, boisterous places. Gulls depart and arrive on constant journeys to feed on neighbouring farmland, where springtime ploughing provides a rich diet of worms and insects. Potential predators visiting the colony are greeted with an eruption of birds into the air and the noise can be deafening. It is clear that the gulls derive considerable protection from the principle of safety in numbers. There are drawbacks to this strategy however. Young birds tend to wander from their nest areas and are often pecked to death by neighbouring parents. (This also happens to young duck and Moorhen attempting to make their way through the colony to the shore) In dense vegetation this may happen less frequently, but the young gulls are also very susceptible to cold and damp, and heavy rain results in many casualties in the long wet grasses. In spite of these threats the species is clearly thriving on the Lough, and towards the end of each nesting season large numbers of young birds can be seen swimming cautiously round their islands or flying weakly out over the shores.

Herring, Great and Lesser Black-backed Gulls nest widely about the Lough, in mixed colonies. Often the establishment of these very aggressive birds is detrimental to other species on the island, which are either prevented from nesting or lose their eggs or young from predation if they do succeed. In common with the rest of the British Isles, the Lough experienced a large increase in their numbers, particularly of Herring Gulls with over two thousand nesting in 1986. This national trend appears to have ended recently, and in many areas the populations are now declining.

The increased numbers throughout the sixties and seventies put considerable pressure on a number of the Lough's islands. Bird Island off Kircubbin for example, had a well mixed population of wildfowl, waders and terns until it was taken over completely by Herring Gulls, although since then Cormorants have managed to get established. Substantial gull colonies occur on all the larger islands in the middle of the Lough, on the Boretree Islands, and currently they are attempting to establish a foothold on Jackdaw and Dunnyneill Islands. Lately however, the population has shown signs of following the general decline, and this may be partly due to botulism, a disease present in the soil and in garbage, especially in hot weather. During warm spells it is distressing to see these birds collapsing and dying out on the islands, but with such large numbers there seems to be little that can be done for them, although on occasions individuals can be nursed back to health.

The 'black-backed' gulls are more restricted in their choice of nesting islands. Greater Black-backs are largely confined to the Minnis Islands where about fifty nest, whilst the Lesser Black-backs (*Larus fuscus*), the most aggressive of the gull species, nest in relatively small numbers on Boretree and Parton Islands. It is the only species to migrate regularly, overwintering in Spain, Portugal and north west Africa, although more are now overwintering in Britain.

The rarest of all the gulls nesting on the Lough is the inappropriately named Common Gull (*Larus canus*), although its numbers show signs of steadily increasing. The North American name 'Mew Gull' is much more suitable, since its high pitched mewing can often be heard amongst the clamour of other birds before it is actually seen. They seem to prefer the more sheltered islands in the Lough; there is a stable colony of about twenty nests on Lythe Rock, whilst recent rat control has made Drummond

Common Gulls are the rarest nesting gull in the Lough, but their numbers are increasing, particularly in the western islands (C.D.D.).

Island more attractive to the species. The nests are very similar to those of the other gulls, but they seem to prefer to lodge their nests between several larger rocks, often very close to the strandline.

Waders; Camouflage, Distraction and Mobility

All the waders breeding on the Lough - Oystercatcher, Ringed Plover, Redshank and Lapwing - are present in winter; however, they may not neccessarily be the same individuals. In common with the much wider shift of birds northwards in spring, many of the waders found breeding in the Lough have overwintered further south, whilst the Lough's own overwintering waders may be nesting further north in Iceland, the Northern Isles, or Scandinavia.

So far, the nesting season has been dominated by species that have devoted time and energy into the creation of nest structures. The waders, and also the terns, employ a completely different strategy. Perhaps because they nest later, in milder conditions, or because increased food availability allows them to spend more time guarding and tending the young, little or no effort goes into nest construction, and the eggs are normally laid out in the open, apparently accessible from all directions. Instead, success relies considerably on camouflage. Laid on a variety of substrates, varying from short grasses, strandline vegetation, dead seaweed and storm debris, or just on shingle, stones or sand, the nests are surprisingly difficult to see. All the eggs are decorated with a variety of blotches, spots and dashes that break up their outline very effectively. The colour varies from a creamy white with dark brown dashes on Oystercatcher eggs (matching the colour pattern of shell debris), to the tiny speckles on the Ringed Plover eggs (they prefer fine shingle or coarse sand) or the darker eggs of the Redshank. Naturally, these patterns of camouflage do not work perfectly for every nest on every substrate that the birds may nest on, but they probably work in the majority of cases. Certainly nesting conditions on Strangford's islands appear to allow for very effective camouflage, since many nests are extremely difficult for humans to see, and we may infer that potential predators are also often beaten by this strategy.

The first signs of waders nesting are small trial scrapes in the shingle on island beaches and spits. Usually the Oystercatchers are the first to start this behaviour, creating one or more small depressions (easily confused with human footprints) where the male has rotated, crouched on his breast, kicking out stones and shells with his legs. The female then selects which scrape she will lay in. Later, Ringed Plovers indulge in the same behaviour, though the scrapes are much smaller.

Distributions of both Ringed Plovers and Oystercatchers, the most numerous of the Lough's nesting waders, reflect the differences in their

An Oystercatcher nest in the Boretree Islands (W.McA).

choice of habitats. Ringed Plovers prefer the finest materials, with well broken shell debris. They are most often found on the small islands, with well formed spits and bars of light shingle, usually thrown up after the winter storms, at the sheltered end away from the large boulders. The heavily eroded islands off the east shore of the Lough, together with Dunnyneill and Jackdaw Islands, are their main strongholds. Oystercatchers are less selective in their choice of nesting sites, and they are therefore more widespread. Suitable habitats include bare rock, stones, grasses, and nests are even occasionally hidden under Nettles. They too seem to prefer the smaller islands, but it is possible that their reluctance to nest on the larger islands is due to the presence of rats. Although normally solitary nesters, they occasionally seem to be almost colonial - in one year the tiny south-west Sheelah Island held ten nests, all within a few metres of each other.

In many cases the nesting density may be determined by the availability of food nearby. Better food stocks would be able to support more pairs of birds, and it is possible that their territorial disputes when establishing nest sites take this into account.

Redshank and Lapwing only nest occasionally round the Lough and on the islands, although Redshank apparently nested quite regularly in the 1940s on the Greyabbey islands. Lapwings sometimes nest on some of the larger islands, and on local farmland. Often the nests occur between the furrows of ploughed fields around Greyabbey and Newtownards, and in less disturbed areas of low lying meadowland in the south of the Lough. Their presence is easily seen in the form of spectacular aerial displays in which they fly up high into the air and then plummet towards the nest site on the ground, only pulling out of the dive at the very last moment.

Whilst it is often difficult to find wader nests because of their effective camouflage, the presence of a nest or young in the vicinity, is often revealed by the behaviour of the adults. Intruding into a nest area causes the birds real distress, and they employ a variety of techniques to get the intruder, human or animal, to leave. Ringed Plovers wander further down the shore, twittering, limping and dragging a wing, and by feigning injury in this way hope to divert the potential predator away from the nest. Oystercatchers do this occasionally, but more frequently they will fly low over the nest area, calling very loudly and repeatedly, until they are confident the intruder has left. Such nesting pairs can easily be distinguished from the more casual twittering departures of the non-breeding Oystcatchers lingering around the islands.

After an incubation period of about three weeks the young Oystercatchers and Ringed Plovers start chipping at their shells, creating a tiny

Masters of camouflage; a Ringed Plover's nest at Kilcool, Co. Wicklow. In the Lough they nest on the spits of finest shingle.

hole through which the young bird's bill can be seen working away to crack the shell. When they finally emerge, young waders are not the naked, defenceless little creatures that are found in the nests of garden birds. They can walk and even run within a short period of hatching. Young Redshank are particularly mobile, and have amazingly long legs for their size which must help them through long grass and strandline vegetation.

The mobility of these young waders poses a potential problem for the parents, since the young birds frequently wander out of the patch of ground that has been defended as the nesting territory. The problem is solved by the adults transferring their territorial behaviour to a 'chick zone' around the young birds. As the chicks wander, so the protective instincts of their parents adjust so that there is always a patch of ground around the young birds where the parents will, if provoked, go into the regular defence and display behaviour. If the chicks become separated, the parents may have to defend several patches of ground at one time. For much of the time however, and certainly when there are human intruders present, the young birds will remain still, crouched low to the ground. In this position they are, once again, extremely well camouflaged, blending with the surrounding stones and shore debris.

At this stage the parents spend less time sorting out territorial prob-

Sandwich Terns on Jackdaw Island. In some years over a thousand pairs of this unpredictable species have nested here, and in others, virtually none (W.McA).

lems. This is because they are now likely to be nesting at the most appropriate density for the local food supply. Once the nesting density has been 'organised' in this way, and and the young hatched on a particular island, further disputes are unneccessary. This allows the parents to spend more time feeding the young, and keeping a wary eye open for predators.

Terns – Safety in Numbers

By contrast with the waders the terns are truly colonial nesters, and on the Lough, spectacularly so. In 1984, the All Ireland Tern Survey, conducted jointly by the Royal Society for the Protection of Birds and the Irish Wildbirds Conservancy, found that nearly one third of all the terns breeding in Ireland did so on Strangford Lough. Almost thirty islands have been used by these active birds in the last twenty years, with varying degrees of regularity. The colonies range in size from a handful of scattered nests which the birds have to work constantly to defend, to densely packed areas holding as much as 1500 nests protected by the intimidating presence of perhaps as many as 2500 birds at any one time.

Tern colonies can usually be heard from a distance. Constant disputes over nesting sites result in an endless series of raucous calls that rises to a deafening crescendo with the arrival of a potential predator. Birds are constantly leaving or returning to the colony with Sand Eels or Herring fry. Out on the open waters of the Lough, the presence of rich plankton and its attendant shoals of small fish forced to the surface by turbulence, result in dense flocks of feeding terns. Hovering kestrel-like over the restless water, they dip and half-dive, and skim back and forth over the surface for a suitable target before finally diving for the tiny fish.

Four species of tern have nested on the Lough in recent years, and because their nesting sites are often under threat from tourism and other coastal developments, a close watch is kept on their fortunes on the Lough. Sandwich Terns, the largest of the species and easily distinguished by the pronounced crest of black feathers and the heavy black bill with a yellow tip, are the first to arrive and the first to nest, usually beginning in late May. They first nested in County Down in 1906. At present they nest on relatively few islands in the Lough, the most notable ones being Ogilby, the Sheelahs, Jackdaw and Swan, but these are large tern colonies, the Jackdaw Island colony for example, having held over a thousand nests in some years. Nesting is dense; many nests are well within a foot of their neighbours, and the ground soon becomes caked in their droppings. Their numbers have shown a general increase in recent years with total counts frequently exceeding two thousand nests.

Common and Arctic Terns begin to arrive shortly after the Sandwich, and they nest slightly later. Both species are smaller and lighter than

A Common Tern with young on Ogilby Island (L.J.T.).

Sandwich Terns, and have long swallow-like tail streamers. Collectively they are the most widespread of the nesting terns in the Lough, and although they do form single-species colonies, often they mingle together, and since both the birds and their nests are so similar to each other, finding out how each species is faring requires a lot of careful observation. At least 28 islands have been used by these birds in the last twenty years, the colonies ranging in size from a few nests to more than 260 tightly packed along an island fringe. Both species appear to be maintaining their numbers on the Lough, Common Terns usually with over 500 nests, Arctic Terns with about 200 nests each year.

Finally, the Roseate Terns (*Sterna dougallii*) arrive ....or should do. These are the rarest of all the nesting birds, and in common with most of north west Europe, their numbers on the Lough have declined rapidly. In 1982 forty pairs nested on the north side of Dunnyneill Island; and in 1985 a similar number nested on Jackdaw Island, but since 1986 none have nested in the Lough, although occasionally they nest on Sandy Island near Ballyhornan at the entrance of the Lough. Elsewhere in Britain and Ireland, the picture is no better with colonies declining or disappearing entirely. The suggested reasons for their decline have included failure of their winter food stocks, trapping by west African children for fun and food, long-term shifts in their breeding range, and disturbance of their European breeding grounds, mainly by summer trippers. Although disturbance has undoubtedly been a problem, the widespread decline of Roseate Terns is probably due to more fundamental changes taking place in their lifestyle, migration, feeding, or general biology.

As if the situation was not complicated enough, terns, as a group of species, are notoriously fickle in their choice of nesting sites. There have been many instances where a successful site has been abandoned without any obvious reason. This has occured a number of times on the Lough, with abrupt shifts of several hundred pairs between the Boretree Islands and Ogilby, and from Jackdaw Island to Dunnyneill, Gabbock and the Sheelah Islands. In 1989 some thousand pairs of Sandwich Terns deserted Jackdaw Island, although this loss was compensated for by increases elsewhere in the Lough. This behaviour makes the task of looking after their interests very difficult, and the general strategy has wisely been to ensure the availability of as wide a selection of islands as possible from which the birds might choose.

All four species perform similar courtship rituals. Some may form pairs as early as February, whilst they are in Ghana or Sierra Leone, but many select a mate after they arrive in the Lough. Courtship, with one or more males vying for the attentions of a female, can easily be seen taking

Roseate Terns on Rockabill, near Dublin; declining throughout Europe, the species has almost completely disappeared from the Lough.

place over the colony site. The most accessible place to see this is from the quay in Strangford village harbour, where Sandwich, Common and Arctic Terns all nest on Swan Island. Two or three birds fly high over the island, and then chase each other down, weaving back and forth, their wings held in a tight 'V' shape over their backs, releasing the air to descend in a steep glide, calling to each other incessantly. Picking up speed, they dart over the island and in between the yacht masts, pulling out from the dive and streaking out over the Narrows. It is a spectacular sight throughout the breeding season, particularly in courtship before nesting commences, but one is tempted to think that it is sometimes done out of sheer fun.

Later, as pairs become established, the male will return several times with small fish or shrimps for his mate. Females have even been known to beg for a 'gift' before they will consent to copulation with the male. Like the waders, male terns make several scrapes to entice the female. They do this by crouching low, breast pressed close to the ground, kicking out stones and sand in a tight circle with their legs. On some of the more rocky islands this must be a somewhat futile operation, but on the spits and bars so characteristic of the Lough's islands, the fine shingle and shell debris is ideal, and a number of well formed scrapes are produced, one of which will be finally selcted by the female for laying.

In spite of this rather energetic courtship and selection of nest scrapes, the tern's final nesting efforts are minimal by contrast with those of wildfowl or Cormorants. Sandwich Terns usually nest a few yards inland from the shore (sometimes much further - one colony on Jackdaw Island was high on the centre of the island, about a hundred metres from the shore) on short grass, or higher strandline vegetation, and islands like Jackdaw or Ogilby are ideal for this. In such areas the scrapes are slight affairs, and it often seems as though the eggs might roll away. Both Common and Arctic Terns are much less selective in their sites as well as their islands. On Swan Island they nest in long grass and nettles, whilst on Green Island rock, Gabbock Island, and the Chandries they nest on the dead seaweed precariously close to the high water line. These banks of dark, crisp-dry weed and gravel can become surprisingly warm - one such site on a warm July day reached a temperature of 39°C suggesting that on occasions the birds have to incubate to keep their eggs cool, as well as to keep them warm in more normal temperatures. In dull or wet weather, prolonged absence from the nests can chill the eggs or young resulting in death, a good reason why their colonies should be left undisturbed. Moreover, the occurrence of spring tides, especially if they are backed up by southerly gales, can lift the weed banks and their nests clear of the island, while the frantic parents fly helplessly over their floating eggs. Such disasters may result in the parents laying a second time if it is not too late in the season, but the clutches may be smaller in such cases and have less chance of success.

Roseate Terns on the Lough have always tended to be secretive in their nesting. They are the last of the tern species to nest, and by the time they are ready to select their nest sites, the island vegetation has grown quite considerably. Frequently they select well hidden spots under clumps of Sea Mayweed, Prostrate Orache, or Sea Campion. The eggs are more elongated than those of the other species, and being darker and more finely speckled are well camouflaged in the undergrowth. All this has made their numbers, and their decline, even more difficult to monitor, although their presence is usually revealed by the much hoarser call than those of the other species.

All the tern species seem to derive considerable benefit from nesting close to the more aggressive Black-headed Gulls. Although there must be losses from egg and chick predation by the gulls, this appears to be offset by living next to large numbers of a vigorous species that actively keeps predators at bay. Where rats have been eliminated from an island, the Black-headed Gulls are often the first to respond to the more favourable nesting conditions and the terns are not far behind, coming to nest under the 'umbrella' of the gulls' aerial defenses. In spite of this, some predators can exert a heavy toll. Visits by family parties of Hooded Crows have resulted in the destruction of several hundred eggs in a single morning. Peregrines may approach unseen, low over the water, rising to swoop over the island at the last moment catching the colony by surprise. Fishing expeditions may be fraught due to the piratical attacks of Arctic Skuas (*Stercorarius parasiticus*) which select a bird and chase it until the terrified victim drops its catch or vomits up its last meal, which then is eaten by the Skua.

Incubation of the eggs continues for slightly over three weeks. At the end of this period the young start chipping, using the white egg tooth at the tip of the beak which they will retain for their first few weeks of life. Gradually the small hole in the shell is enlarged, and it is possible to see the egg tooth working back and forth within the shell. Occasionally the young birds can be heard cheeping to their parents from within. They hatch helpless, wet and bedraggled, but soon dry off and within a few days are able to support themselves within the nest. After hatching the parents transfer their routine of returning with fish for each other to one of feeding the young. This is the most energetic phase of the whole season; a constant series of expeditions for both parents to areas rich in fish, whilst at the same time defending their vulnerable young from predators and territorial disputes. Over forty meals must be provided daily for two chicks in the nest, and with that demand, it is clear that the proximity of abundant stocks of small fish plays an enormous role in helping the birds to meet that challenge.

As the season progresses and the demands of growing chicks become ever greater, Sandwich Terns, in particular, habitually take short cuts across land to get to and from other fishing grounds. Thus birds nesting on Ogilby and the Boretree Islands can often be seen crossing the Ards Peninsula over Greyabbey and Mount Stewart to fish off the Portavogie islands, whilst Jackdaw and Swan Island birds cut across headlands like Killard Point, Kilclief, and right over Strangford village, the Black Causeway, and Castle Ward. They also commute to Belfast Lough.

The young birds grow rapidly on their diet of fish - within three weeks they are running around their island. A successful season for Sandwich Terns will eventually result in hundreds of chicks running along the shore of the island in 'herds,' flexing their developing wings as they run. Within a few days these young birds will be half flying, half stumbling over the boulders and weeds of their island birthplace, in their first efforts to acquire the flying skills that will shortly be needed to take them back to Africa.

Outside the Lough

Finally, it is worth spending a few moments considering some of the birds nesting outside the Lough, since many of them are regular visitors to the Lough at other times of the year or in poor weather. Closest to hand, immediately to the southwest of Killard Point, lies Gun's Island. A large wedge-shaped island, its high outer shore terminates abruptly in a complex of cliffs and low crags. There is a rapidly developing Kittiwake colony here, their nests perched precariously on small ledges in the rocks. In recent years they have increased to over 140 nests, and the large numbers of nonbreeding yearlings associated with the colony suggests that future years may see a considerable expansion in the colony, possibly towards the cliffs at Sheeplands. They, and the dozen or so pairs of Shag, probably commute regularly into the Lough, the Kittiwakes often being seen in the Angus Rock area. Around the rest of Gun's Island nesting is dominated by about two thousand pairs of Herring Gulls and a few Great Black-backed Gulls. The most sheltered portion of the island trails off into a spit terminated by a tiny little slip of land covered in sand and fine gravel, colonised by Sea Mayweed and Prostrate Orache. This is Sandy Island, and although it has suffered a variety of disasters, including incursions by cattle, Foxes, rats, humans and spring tides, it sometimes supports a vigorous tern colony of well over a hundred Common and Arctic Terns and often one or more pairs of Roseate Terns.

To the north of the Lough's entrance, there are a number of small islands, all rather bleak, and in winter exposed to the full fury of Irish Sea storms. The South Rock lighthouse supports a small colony of Shags (the rock itself is virtually submerged at high water), whilst the North Rock is colonised by some seventy pairs of Cormorants in uneasy association with about fifty pairs of Herring Gulls. Burial Island is almost smothered by over a thousand Herring and Greater Black-backed Gull nests, but even in this unpromising situation it appears that Cormorants may be gaining a foothold with the recent establishment of a small colony of over twenty nests.

The Copeland Islands also hold major colonies of nesting birds in addition to the Eider Duck mentioned earlier. Most significant are the populations of Manx Shearwaters, with up to a thousand pairs burrowing into the thin soils of Lighthouse Island, and about a hundred on Mew Island. One bird, ringed by the Copeland Bird Observatory in 1953 was retrapped in 1989 - a lifespan of at the very least, thirty-five years. These graceful, strong flying birds are equally at home as they sweep over the waters off Rio de Janeiro, the Azores, or off Ballyquintin Point, where sometimes they can be seen in rafts of up to 2000 birds, attracted into the

Black Guillemots ('Tysties') hunting for Butterfish at Marlfield. They willingly take to nest boxes provided by the National Trust.

area by good fishing. Other species include over forty pairs of Common Gull, a growing and vigorous colony of Arctic Terns on Copeland Island itself, a small but regular number of Black Guillemots (known as 'Tysties' in Scotland) on Lighthouse and Mew Islands, together with a few Oystercatchers and Ringed Plovers.

Closer in to the shores of the Ards Peninsula, Bird Island, off Portavogie usually has a substantial colony of about a hundred mixed Common and Arctic Terns, whilst Cockle Island, in the middle of Groomsport Harbour supports an Arctic Tern colony with over 100 nests. It is likely that individuals and pairs from many of these outlying colonies interchange with populations in the Lough. This is particularly likely to be the case with the terns, with their habit of abruptly deserting one colony site in favour of another. The fact that terns regularly commute across the Ards Peninsula in both directions suggests that all the colonies may be linked in this way, and possibly interchange with other sites much further afield.

Black Guillemots are found on the Outer Ards Peninsula, and in the south of the Lough. This is very much a northern species, regarded by southern bird watchers as an exciting rarity. These small black and white auks, with their brilliant red legs and mouths are becoming an increasingly frequent sight in local harbours and inlets. Their natural nest sites in cliff holes and crevices are scarce, but welcome artificial substitutes have been provided in the form of special holes in Bangor pier, whilst in the Lough these adaptable little birds have taken willingly to nest boxes provided for them by The National Trust. Such sites now regularly have small rafts of birds fishing, especially for Butterfish, in the waters nearby.

The End of the Season

As the nesting season draws to a close, the fortunes of the season become apparent. From May onwards successive goups of young birds make their appearance. Close-knit family parties of Mallard and Canada Geese venture out on to the open waters of the Lough, the young birds often vigorously battling against difficult wave conditions. Brown speckled young gulls flap weakly over island shores, their calls sounding like squeaks by contrast with those of their parents. Many young birds never get this far. Large colonies may often be littered with carcasses - the victims of cold weather, territorial disputes, death of the parents, food shortage and predation.

For the survivors, there will be a period of a few months when the Lough's waters will be at their warmest. Food, whether it be fish or invertebrates, is plentiful. This is a time of consolidation, when the young

Sunrise over the Narrows. Nesting birds, wintering wildfowl and the rich array of marine life all depend on its powerful tidal currents (A.J.).

CONSERVATION OF NESTING BIRDS

Nesting birds require the right habitat, nesting materials, and freedom from predators and human disturbance, in addition to adequate food resources nearby or further afield. Some fifty islands are important for various types of nesting birds, and many are managed by the National Trust for this purpose, either through ownership or by agreement with owners.

Early in each year Rats are erradicated; they can cause extensive losses of eggs and chicks. Vegetation on tern nesting areas is either cut or sprayed to keep it short for them. Elsewhere, pond improvement and planting of scrub, sometimes with the aid of local wildfowlers, improves the habitat for nesting duck. On islands particularly important for sensitive species like terns, aggressive species like Hooded Crows, Herring and Greater Black-backed gulls are prevented from nesting because of the damage they do.

Human visitors to nesting islands can wreck nesting, through disturbance and even unwitting trampling of nests. Therefore nesting islands are marked with signs requesting visitors not to land. All signs are removed after the season has finished. There are however, many other islands where no sensitive species are nesting, and where it can, the Trust actively encourages people to visit these to picnic and enjoy the scenery and wildlife. Support for this policy from the boating community of the Lough has been excellent.

---

birds put on weight and gain strength. Young gulls float with the adults in enormous late-summer rafts on the calm waters near plankton and food-rich upwellings in the south of the Lough. Such areas act as magnets for large numbers of both adult and juvenile terns, often darting and diving in dense flocks around the gulls, particularly in the Narrows. Oystercatchers remain with their parents well into September, still receiving some assistance with the task of cracking open Cockles and Shore Mussels. Young Cormorants fish for themselves, but remain faithful to the colony for a while, although later many may fly to other parts of Ireland including Lough Neagh, or to south west England, Brittany, or down to the Bay of Biscay. Other young birds come into the Lough from further afield. Young Razorbills and Guillemots for example, can often be seen swimming closely behind one of their parents, having swum from colonies on the Calf of Man, Muck Island, and the north Irish coast.

Young terns have much less time to develop and mature. Within a few weeks of fledging, many are already on their way south. Young Sandwich Terns may linger on for a while, following their parents around the

Lough. They and their parents (now crowned with the white flecks and patches of winter plumage) can often be seen roosting on beaches and sand bars, all facing into the first September breezes, before they finally depart for west Africa. Many seem to fly in groups; flocks of terns from islands in Strangford Lough have all been known to arrive at sites in north Wales within a few hours of each other, but at present we are uncertain if they stay together all the way to their wintering grounds.

Whilst all these young birds are learning to cope with the realities of life on the wing, a gradual change comes over the Lough. The fresh greens and golds of spring on the islands and drumlins around the shores give way to the dark greens and weathered browns of high summer. There are new sounds on the Lough as well. Even as early as June the first Curlew, probably nonbreeders from moorland Britain, are returning often closely followed by Lapwing. In August the first Dunlin returning from their breeding grounds on the machair of the Hebrides, still bearing the russet colours and black breast plumage of summer, appear on the Lough. Within a couple of weeks, the first Brent geese return from Arctic Canada. Then the family parties make their appearance, the young readily distinguished by their pale wing bars. For a while the young Brent and the young terns - representatives of the Arctic and the Tropics - share the Lough, as their parents did in early spring. Then just as in the spring, they part company, this time the terns embarking on their long journey to the rich winter fishing grounds off the west coast of Africa.

9   Postscript

NOBODY KNOWS how many different wildlife habitats there are in the Lough, or around it. However we do know that there are many. Every boulder for example, has its underside, sunny side, sheltered side and exposed side, with a further selection of conditions to be found in the pits and cracks of the surface. This applies to all other features of the Lough. At a larger scale however, there are major habitats, both widespread and obvious to human eyes. The differences between a saltmarsh and a rocky coast are readily appreciated, as are those between the current-swept bottom of the Narrows and the fine silts further into the Lough.

Viewed from the perspective of the wildlife, matters are not so straightforward. Every plant or animal has its own preferred range of conditions in which it lives. Each responds to individual factors, like salinity, temperature, nutrient content, water movement, or the texture of the land or sea bottom, as well as to their combined effects, and other features of the Lough like food supply.

Some tolerant species, like the Common Shore Crab, are widespread throughout a variety of conditions and habitats. By contrast, the tiny Pea Crabs (*Pinnotheres pisum*) whose whole adult life is spent feeding and breeding inside Horse Mussels and other shellfish, are so selective one may be forgiven for wondering how they survive at all. Others, like the beautiful Sea Pinks, can be seen blooming on the most sheltered salt marshes and at the same time out on the most exposed rocky headlands; further inland however, they are very rare.

Such inconsistencies, as we see them, in their choices of habitat and conditions, are almost certainly quite logical responses to a whole range of factors as experienced by the plants and animals themselves.

Taken a stage further, plants and animals in responding variously to these stimuli, come together as assemblages of species in particular areas. But they are not just neighbours; they interact with each other in the complex web of relationships we examined in the preceding chapters. The plants harness the sun's energy, provide gaseous exchange, and much of the physical structure of some areas, whilst the animals perform roles of grazers, filterers, predators, parasites, defecators, pollinators and transporters. In this way communities are formed, and we can see the way these change from one area to another. But once again, we cannot leave it here.

Such communities do not operate on their own. They interact with each other. In part they do this by modifying conditions in the whole Lough; for example by producing or consuming plankton and nutrients, or by changing physical conditions, in the way that a salt marsh develops on a shore. They also interact through the mobility of their wildlife. Plants disperse offspring and fragments of themselves either to colonize or to be consumed in other areas and communities. Animals move around, and some predators clearly interact with a large number of communities. In this sense then, there is no such thing as a truly distinct community since they are all interconnected. We may be tempted to think that the most obvious divide of all, that between the sea and the land, separates communities. Even this is not true, for there are powerful links through the flow of nutrients, erosion, action of seabirds, and the shelter and consolidation that vegetation can provide. Thus the Lough acts not as a loose cluster of habitats, species, or communities, but as a single system operating at many different levels, whose unique complexity and wealth are the subject of this book.

Can we at least draw a line at the entrance of the Lough? The simple

A Greenshank in Castle Ward Bay; one of many predators and just one part of the Lough system.

answer is no; the links with the Irish Sea through the Narrows are strong ones, not only in terms of water movement and water quality, but also in terms of the interchange of marine life between the two areas. Strangford Lough, by virtue of its productivity, may well enhance conditions in the Irish Sea outside. Ultimately, we must appreciate that all seas are interconnected; there is only one ocean, and Strangford Lough is a small but intrinsic part of it. Moreover, the massive movements of bird life on long distance migrations reinforce the Lough's global role, and as such it is dependent on the fortunes of these birds in the arctic and the tropics, whilst its own resources contribute substantially to their success in those areas.

The Lough system is not static. Natural variations in the fortunes of various species, slight shifts in the direction of prevailing winds, are continuously altering the distribution of habitats, and in some cases eliminating them altogether. On a longer time scale, it is almost certain that our climate and sea levels are still changing in association with the profound effects of the Great Ice Age. Moreover, geologists tell us that the average life of a species is only about three million years; all the species in the Lough and elsewhere are therefore undergoing their own changes. Natural change is thus built into the whole system, but like a well oiled machine, all the components have the capacity to adjust and respond, giving an overall stability to the Lough.

Just Another Species...?

So far, human activities have not featured significantly in this account of the Lough's wildlife, and it is now time to consider these. The earliest settlers arrived at a time when many of the wildlife communities and their various relationships were becoming established after the retreat of the glaciers. The mild climate, resources of edible wildlife and relatively sheltered conditions must have been a considerable attraction to those early settlers. They hunted for fish, shellfish, birds, large and small game, and gathered fruit, nuts, and edible plants. They must have exerted some influence on the wildlife of the Lough, as all species do, but it was probably minimal. Excessive depletions of any of the important resources would almost certainly have forced some restraint on those first settlers. In other words, they were *part* of the complex Lough system, and for their own survival needed to live in harmony with it.

Since those early beginnings, the development of technology gradually increased man's ability to exploit the Lough and change its environment and wildlife. Early agriculture resulted in small scale temporary clearings, but with improved equipment and population increases over some 4000 years, these became extensive and permanent, and it is likely that such areas experienced increased erosion. By early Christian times much of the country was colonized in this way, as the distribution of some 1300 raths (farmsteads) in County Down testifies. Some thousand years later, the final removal of most of the Lough's woodland, left us with today's landscape of hedges and small fields. At the same time marshes were drained and land fertilized. Thus a number of habitats and communities with their associated species had been altered or destroyed completely.

In the waters and on the bed of the Lough, developments proceeded more slowly. However by mediaeval times fish traps had been constructed, and there was probably considerable use of seaweed for fertilizer, and removal of gravel and sand for building. The development of efficient and manoeuvrable boats for exploiting deeper resources took longer. But the effects of human activities were beginning to spread to the waters in other ways. Growth of towns and villages increased the flow of sewerage, and therefore the input of nutrients into the marine system; shore was being claimed for agricultural land by the construction of tidal banks, whilst causeways were built to link islands formerly separated by tidal channels. All these projects influenced the patterns of sediment and water movement about the Lough, and continue to do so in a way that even now we barely understand.

These were the first indications of a trend that is now firmly established: human abilities to change conditions on the Lough through exploitation and development considerably exceed the understanding of what their consequences are for the system of the Lough. Initially, due to relatively simple technologies and lower population, these changes may have been sustainable to some extent, but today this situation has changed dramatically.

The last half century has seen an enormous expansion in the influence of man on the Lough system. There are now no habitats or communities which are not to some extent affected by human activities. Wind blowing over the Lough carries particles and chemicals released from human population centres, often far away. Water flowing to every submarine crevice carries traces of effluents of many types. Particular species have been selectively exploited or removed, and some habitats, scenery, and in places, topography, have been greatly altered. Moreover, unlike most other species, man's demands and activities have shown no signs of reaching a state of equilibrium with other natural features of the Lough. The changes brought about increasingly tax the ability of many species to survive in the Lough, and in some cases this may threaten whole communities. Thus human developments about the shores and in the waters now have the capacity irreversibly to damage the sensitive ecologi-

Postscript

cal balance of the natural communities, far in excess of those changes caused by most natural biological and climatic changes.

A few examples may illustrate the point. Improvements in boat and gear design, have all served to increase catches of many commercial species in the Lough's waters. A important Herring fishery survived declining catches in the early part of the century because of these advances, but it eventually ceased. Later, small locally based vessels operated an apparently sustainable fishery dredging for Scallops. Its demise coincided with the arrival in the late 1970's of more powerful vessels mainly from outside the Lough, with much heavier gear, and the result was not only destruction of the Scallop population, but also the Scallop habitat. With its decline, vessels turned to trawling for Queen Scallops ("Queenies"), and again, the trawls are also destroying the Queenie's habitat, the extremely rich Horse Mussel community in the centre of the Lough. This productive community, dependent on stable conditions, can only recover very slowly, and its continuing destruction has implications not only for the fishery, but for the sediments, for the plankton content of the Lough, and also for the dense populations of about a hundred species that form the community. Failure of Queenie fisheries has already been reported for some areas off the Scottish coast; investigations have revealed that substantial areas of the Lough's Horse Mussel community have already been lost.

Trawlers fishing for Queenies on the Horse Mussel beds. Lack of fisheries management has been detrimental to both fisheries and wildlife.

Although biological factors may have played a role in the fluctuations of some of these species, the repeated cycles of abundance at the beginning of a fishery, followed by a period of shortage, and eventual closure of the fishery, point to a profound imbalance between the fishery techniques applied and the ability of the Lough to sustain these demands. In human terms this shows a lack of proper regulation to the detriment of both fisheries and wildlife. Destruction of the habitat of a species important to a commercial fishery must surely represent one of the most shortsighted actions to be undertaken by any industry.

Other methods of harvesting food from the Lough's waters are less destructive. Potting for Scampi, crabs and Lobsters, and diving for Scallops, takes the target species, but can be achieved whilst leaving the bottom communities largely unaffected. Oyster cultivation (*Crassostrea gigas*) can be practised in some areas, and provides a strong political impetus for good water quality. Currently experimental Scallop cultivation is taking place, giving rise to the possibility of regenerating the wild population from spawning by the farmed stock. Both oysters and Scallops feed naturally during cultivation by filtering plankton and other microscopic material from the water.

Very different is the operation of fish farming, which requires the input of very large quantities of pellet food into cages of crowded fish. Much of this, together with faeces, falls uneaten to accumulate and rot on the bottom, putting the ability of the natural wildlife community to absorb this deposition under grave stress. With their organic pollution, highly toxic fish medicines, shooting of seals and fish-eating birds, and hazards to navigation, such developments are already the subject of concern in many other areas, and would certainly be incompatible with both conservation and amenity interests on Strangford Lough.

Another potential threat is the uncontrolled increase in the human population of the area. Increases in the size of the loughside towns and villages could result in increased sewerage production; whilst most of the Lough appears to be absorbing the present supply of nutrients, we cannot rely on the tidal currents to remove the effects of any marked increase. Some areas in the north of the Lough may already be showing signs of nutrient enrichment, whilst the natural communities of small bays and inlets are already suffering from excessive bacterial growth caused by farm effluents.

Increasing use and development of the Lough, partly through leisure and improved communications, is already changing its character. Ill-considered and insensitively sited developments are eroding the natural landscape and countryside about the Lough. On the shores wildfowling in a number of areas has adversely affected stocks of wintering birds, and this trend is now supplemented by increased numbers of shellfish collectors, horseriders, and other activities. On the waters, the movements of boats in themselves do not generally affect the wildlife, but the unplanned spread of moorings is a major problem for the maintenance of unspoiled scenery, and for marine life in shallow subtidal areas.

These greatly increased pressures pose a major threat to the natural system of the Lough. It is doubtful if it has the capacity to absorb these changes if they continue at their present rate. If the unique balance

---

STRANGFORD LOUGH AS A WILDLIFE RESOURCE

- The majority of major inshore subtidal habitats to be found in the British Isles are represented in the Lough.

- About ten major intertidal habitats are represented in the Lough.

- Over two thousand species of marine animals already recorded, or about 70% of species found around the Province.

- The largest population of Common Seals of any site in Ireland.

- Internationally and Nationally important populations of wintering birds use the Lough.

- The Lough has been designated a Sister Reserve with North Bull, Dublin, and Polar Bear Pass, Canada, on account of its importance for the shared population of Pale Bellied Brent Geese.

- 79% of the bird species found in Northern Ireland in winter have representatives on the Lough.

- About one third of all the terns nesting in Ireland in summer, do so on Strangford's islands.

- The Lough holds important areas of salt marsh; in Northern Ireland, second only to those in Carlingford Lough.

- The Lough has been designated an Area of Outstanding Natural Beauty, and much of it as an Area of Special Scientific Interest. Work is progressing slowly towards its designation as a Ramsar site, and as a Marine Nature Reserve.

- Astonishingly at present there is no coordinated management of the Lough to safeguard these resources, although discussions on the need for this have continued for some 15 years.

between the habitats of the Lough with their rich array of communities and species is to be maintained, we need to find ways of using the resource more wisely. In other words it needs to be conserved. This does not mean preventing any further use of the Lough; rather it is the reverse, seeking to ensure, by careful planning and integrated management, that its resources will still be available for future generations to continue exploiting. Although we do not know everything about the environment of the Lough, we do know enough to achieve this to a large degree, given the willingness to do it.

In 1966, because of the concern at the pace of change felt by local naturalists, scientists, landowners and wildfowlers, the Strangford Lough Wildlife Scheme was established by The National Trust. Although initially controlling wildfowling (through a covenant with the Wildfowler's Association of Great Britain and Ireland, now the British Association of Shooting and Conservation) it expanded to take account of all aspects of the Lough's wildlife, viewing the Lough as a single system. It continues in this way, the Trust devoting substantial resources to the operation. It now manages, through ownership, leases and agreements, most of the shores and a number of islands and areas of mainland. Other bodies have also become involved in the Lough's conservation, notably the Wildfowl and Wetlands Trust, R.S.P.B., the Ulster Wildlife Trust, the Ulster Society for the Protection of the Countryside, the British Trust for Ornithology, the World Wide Fund for Nature, and the Marine Conservation Society.

This returns us to the earlier question: can we draw a line at the entrance of the Lough? The biological answer was no, but the practical management answer is probably yes, though only if it is through a fully co-ordinated and effective structure that recognizes the Lough's relationship with the Irish Sea and its major national and international roles.

The National Trust and all the other organizations concur in their view that the Wildlife Scheme on its own is simply not enough to achieve this. Whilst it may cater for matters specifically associated with wildlife, the lack of a co-ordinated plan for the development, recreation, and scenic value of the Lough means that to a large extent the Scheme is operating in limbo, without any framework in which to plan for wildlife conservation. For many years they have pressed for the wider conservation management of the Lough. An early stage in the process saw the enactment of the Amenity Lands Act of 1965, and in 1985 the Wildlife (N. Ireland) Order, and the Nature Conservation and Amenity Lands Order. Most of Strangford Lough was designated an Area of Outstanding Natural Beauty, and several National Nature Reserves were established. More recently most of the Lough shore has been designated an Area of Special Scientific Interest. Currently discussions are underway about declaring, as the Minister for the Environment put it, 'the entire area of Strangford Lough a Marine Nature Reserve.'

These designations, although welcome and effective to a limited extent, do not resolve the central problem of a piecemeal approach to the managment of a resource that operates as a single, complex unit. Over the last twenty-five years the government has been distinctly reluctant to respond to the pressure for an overall structure for the Lough's management, although there have been considerable efforts on the part of voluntary groups to work in this direction. Until such a framework is provided, catering for both the Lough's internationally important range of wildlife communities and for appropriate, controlled use and development, its long term future remains at risk as the pressures on the system increase.

If we cannot meet the challenge of wisely using an area so rich as Strangford Lough, the future is even more bleak for the much wider areas of countryside and coast that are less conspicuously endowed with natural riches.

There is time yet, but not much. The wildlife of the Lough is still in good heart, and its constantly changing moods are there to be experienced by everyone. The more people get out and enjoy the Lough and its wildlife, the greater the impetus for its conservation. I hope this book helps people to get more interest and excitement from what they see, and I hope it helps to explain why the wildlife is so important and that it must be conserved for the future. Most of all, the book should not become a record of what has been lost.

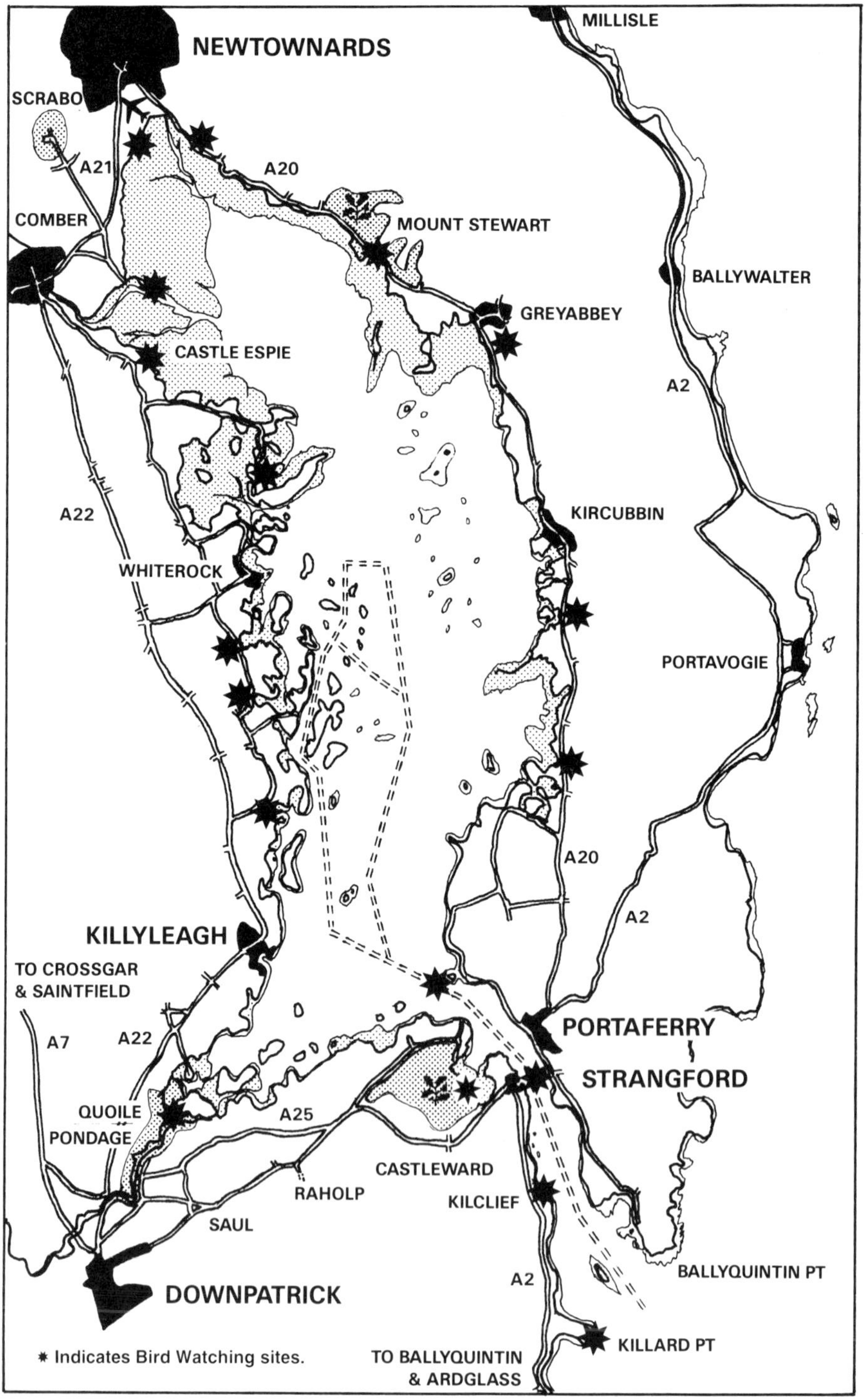
NEWTOWNARDS
MILLISLE
SCRABO
A21
A20
COMBER
MOUNT STEWART
BALLYWALTER
GREYABBEY
CASTLE ESPIE
A2
A22
KIRCUBBIN
WHITEROCK
PORTAVOGIE
A20
A2
KILLYLEAGH
TO CROSSGAR
& SAINTFIELD
PORTAFERRY
A7
A22
STRANGFORD
QUOILE
PONDAGE
A25
CASTLEWARD
RAHOLP
KILCLIEF
SAUL
BALLYQUINTIN PT
A2
DOWNPATRICK
KILLARD PT
✱ Indicates Bird Watching sites.
TO BALLYQUINTIN
& ARDGLASS

Gazetteer

The majority of natural features described in this book are readily accessible to those who wish to find them. Most of the coast, shores and waters are within relatively close range of roads and shoreside towns and villages. The best investment the visitor can make in the process of exploring the Lough's wildlife is time. Time to explore, move around, and search; on other occasions simply time to wait and see what happens - often this is more rewarding.

This is a summary, in two parts, of some of the many places of interest that can be seen around Strangford Lough. It is not practical to include all the locations that could be mentioned, so I have concentrated on those that have particular relevance to matters raised elsewhere in the book. The main gazetteer is provided for those travelling by road, and the assumption is made that visitors are coming from Belfast, starting in Comber, heading north towards Newtownards and then travelling southwards along the eastern side, to go round the Lough in a clockwise direction. Obviously there is no need to follow this procedure directly; the route may be joined and followed at any stage in the circuit. The Northern Ireland Ordnance Survey Map (Sheet 21, Scale 1:50,000) is a useful aid during the journey.

There is also a suggested tour for boat-based visitors to the Lough, including some dive sites; this tour is less structured as it has to be sufficiently adaptable to meet the conditions of of weather and tide. However, it has been compiled with a view to presenting the most easily visited locations for those who may not know the navigational challenges of the Lough particularly well.

The journeys are divided into sections, each indicated by a ***bold italic heading.*** Each heading is then followed by a brief description of the route and a few places or views on the way. Then the main places of interest, indicated in *italics,* are described in more detail. Where particular sites, species, or habitats have been described in detail elsewhere in the book, the relevant page numbers are given. Certain facilities around the Lough have been provided or maintained by particular organizations and these are indicated thus: Ards Borough Council - A.B.C.; Belfast Education and Library Board - B.E.L.B.; Down District Council - D.D.C.; Department of the Environment, (Countryside and Wildlife Branch) - D.o.E.(C.& W.); (Historic Monuments and Buildings Branch) - D.O.E.(H.M.B.); National Trust - N.T.; Royal Society for the Protection of Birds - R.S.P.B.; Ulster Wildlife Trust - U.W.T.; Wildfowl and Wetlands Trust - W.W.T.; Privately owned - P.

Finally there is a list of some books which may be interesting or helpful to those who wish to find out more about the Lough's wildlife.

A TOUR BY ROAD

Getting to Strangford Lough (Comber);
From Belfast, via the A22 through Dundonald.
From Saintfield via the A21.
From Dublin, Newry, Rathfriland, via the A25 to join the A22 at Downpatrick, thence to Comber.
From Newcastle and Dundrum via the A2, to join the A25

Comber to Scrabo Hill: From the main square in Comber follow the A21 (signed to Newtownards). Eventually a dual carriageway will be reached; after about 1km of this, turn left at sign to Scrabo Country

Park (D.O.E.(C.& W.)). Follow this road and the signs, gradually gaining height, to reach the car park below the Golf Club, which is almost at the summit. This road runs along the 'tail' of debris left behind by the glaciers of the Great Ice Age.(p. 22). Between October and December look out for Whooper Swans (p. 152) in the adjacent fields.

Scrabo Country Park: From the car park a brief walk NE reaches Scrabo Tower, with its exhibition on the geology and wildlife of the area; a branch to the right leads to the Quarry. From the Tower there are spectacular views of the Lough, the reclaimed land, the islands and mudflats. In winter, look out for high tide flocks of waders 'swarming' along the shores and over the water. Much of the Lough's ancient history is represented in the geology of this site, particularly in the quarry where the Triassic sandstone beds, igneous dykes and sills (p. 18) may be clearly seen. Many rocks near the summit are heavily scratched, and these are thought to be relics of the scouring by glaciers. Particular birds to look out for are Stonechats, Ravens and Peregrines.

Killynether Wood, containing Beech Hazel and Oak, is reached by taking the path south from the car park.

Scrabo to Newtownards; Return to the A21, turning left for Newtownards. In winter look out for flocks of roosting and feeding waders in nearby fields (p. 163). Close to Newtownards the crag-and-tail shape of Scrabo left by the glaciers is apparent. At the first traffic lights turn right, but first notice straight ahead the newly restored town hall (A.B.C.), built of Triassic desert sandstone some 225 million years old, obtained from Scrabo. Follow signs to Portaferry, on the A20.

Newtownards to Greyabbey; About 1 km of joining the Portaferry Road, Newtownards Airport is passed, and shortly after the road joins the northern mudflats (N.T.) of the Lough, rich in burrowing marine invertebrates, and eelgrasses (p. 105). This area is at its best in autumn or winter, when the bird life is outstanding, and it is one of the first areas selected by overwintering Brent Geese on their arrival from arctic Canada (p. 145). The best times for seeing the birds are roughly two hours before or after high tide (p. 161). There is a selection of small carparks and laybys where the birds can be watched; the main ones are described here.

Tidebank: A small car park (A.B.C.) by the sluice gates at the spot where the road meets the Lough shore. From there it is possible to walk along the tidebank, and see flocks of Brent Geese, Shelduck, and many waders, particularly Knot and Redshank. Both Kestrels and Peregrines can also be seen in the area.

Maltings at Ballyreagh: A small stream runs below the road, often carrying spilt grain from the Maltings on the left. This attracts flocks of wildfowl, including Brent, Pintail, Mallard, Whooper and Mute Swans. Unfortunately the traffic makes it difficult to stop here, but immediately to the south there is a layby (N.T.), which gives good views of the birds and of flocks moving up before a rising tide. Look out for massive flocks of Golden Plovers and Knot (p. 162); they may also be seen roosting on the mudflats to the south.

Butterlump Stone: A large boulder of Scrabo Dolerite (p. 23) carried by glacier and deposited about 1km south of the Maltings. It rests on outcrops of the distinctive red Triassic sandstone. The road south from here continues to run along the post glacial raised beach, and there are many examples of the steep banks of drumlins eroded during those times (p. 26). There are several places where grids of stones are laid across the mudflats; these are early attempts by farmers to cultivate seaweed, which was a scarce commodity in this part of the Lough, and was a valuable fertilizer.

Mount Stewart: (N.T.) The first view of this fine estate is extensive woodland on the left of the road south of Cunningburn. The house and its famous gardens, just over 1km further, are open to visitors from April to October; details available from N.T..

Gasworks: (N.T.) Formerly this supplied the estate with its own gas, but it now forms an attractive picnic area and birdwatching hide. The mudflats in this area are rich in eelgrasses, Cockles and Baltic Tellins. In autumn the hide provides an excellent vantage point for watching Brent Geese as well as Mute and Whooper Swans feeding here, and for watching massive flocks moving about the mudflats and off Chapel Island and the Boretrees group. Occasionally there are Buzzards from woodlands in the estate.

Ann's Point: (N.T.) An area of brackish marshland lying behind a tide bank opposite Mount Stewart's Temple of the Winds. Intersected by drainage channels used by Kingfishers, the area is rich in wildflowers, particularly in midsummer. In winter the area is often used by wildfowl and waders; it is a good site for Jack Snipe, whilst flooding in the adjacent fields attracts Whooper Swans, Teal, and Goldeneye. The upper field to the north of the point, and above the raised beach, was a mesolithic settlement.

Greyabbey: The ruins of the Cistercian abbey (D.o.E.(H.M.B)), founded in 1193, may be reached by turning left at the small roundabout in the centre of the village, going to the top of the street, then turning right, and almost immediately finding a small car park, again on the right.

Greyabbey to Kircubbin; Return to the shore, back through the centre of the village, this time going straight on, and continuing on the A20 to Kircubbin and Portaferry.

Greyabbey Bay: Immediately outside the village, there are two car parks (N.T.) by the bay. On the western side of the bay lie Mid and South Islands. The first car park gives excellent views of a small shingle spit which is used as a roost by numerous wintering wildfowl and waders, including Brent Geese, Shelduck, Bar-tailed Godwit, Knot, Dunlin, Ringed Plovers and Turnstone. Between the two is a lake, the 'Swan Hole' in winter often populated by Pochard, Goldeneye and Tufted Duck. Swans nest there in the spring, amongst the reeds. The second car park ('Sloe Bushes') is better for bird watching at lower stages of the tide - look out for Oystercatchers using their anvils to open Cockles (p. 153). Flocks of Long-tailed Tits often forage amongst the bushes. At the head of the bay there are some small, but relatively unspoiled areas of saltmarsh. Long lines of boulders across the mudflats are thought to be mediaeval fish traps.

Wren's Egg: About 1.5km south of Greyabbey lies another enormous glacial erratic, the name presumably being an example of Greyabbey humour! Since the road is fairly straight, it is possible to pull into the verge and look towards a small area of saltmarsh, often used by roosting birds.

South of this point, the road parts company with the raised beach and bends inland behind the shoreline drumlins, after 2km returning to descend towards Kircubbin Bay and the village.

Kircubbin to the Dorn: Continue through Kircubbin, following the A20 on towards Portaferry. The scenery begins to change at this stage; although the glacial drumlins are present, in places outcrops of bedrock are often seen, giving a more rugged appearance to the landscape (p. 10). Rocky headlands poke out in sharp contrast to the muddy bays sheltered between them.

Doctor's Bay: A sheltered muddy bay, with a developing area of saltmarsh, as shown by the bright green swards of Glasswort (p. 117). On the southern side lies Black Neb, a headland richly covered with lichens, Gorse and small areas of maritime heath (p. 124).

Horse Island: 2km south of Kircubbin lies a small bay with a patch of firm ground for parking on the left of the road. This gives access to Horse Island (N.T.), which can be reached on foot at almost all stages of the tide. There is a rough path going round the whole island. Dense thickets of Gorse provide ideal conditions for small birds. Sharp outcrops of bedrock are covered in lichens and other maritime plants (p. 122). The shores are richly covered with weeds and rockpools and provide the first good example of this habitat (p. 93) on the journey. Teal, Redshank, Greenshank, Turnstone may be seen here, whilst further out it is possible to see various auk species, divers, and in summer, terns (p. 73). Around the more sheltered parts of the island there are well developed areas of salt marsh with a good variety of species like Sea Aster, Sea Pink, and Sea Purslane.

Saltwater Bridge: About 2km south of Horse Island, this small bay has fine areas of salt marsh (p. 117), and as the name suggests, high tides flow under the bridge and up the Blackstaff River. There are views out to the Craiglee Rocks, and beyond to the Mournes. The alternation of rocky outcrops and sheltered, very muddy shores make this a particularly rich area for shore life.

The Dorn: The name of this inlet has been used for the whole of the National Nature Reserve (D.o.E.(C.& W.), & N.T.) whose northern boundary lies a few metres to the south of Horse Island. It is worth stopping to view the area from one of the high points on this part of the road (but be careful of the traffic). To the north in the distance lies Scrabo Hill, and closer to hand, Gransha Point, a tenuous spit of shingle and bedrock extending over 1 km into the Lough. The cluster of the Yellow Rocks lies to the west in the foreground, and beyond a scatter of pladdies terminating in Youran and Needoo. This is an important area for Common Seals (p. 75), and they may be seen basking on the closer rocks during low tide. Small parties of wintering Brent, Teal and Shelduck can often be seen foraging close to the road. To the south, tucked behind Castle Hill, the site of a mediaeval castle and church, lies the Dorn inlet itself, a veritable maze of islands, channels and mudbanks (p. 39).

Continue southwards for about 2km, then take the first turning to the right (Abbacy Road). This runs beside some of the most sheltered parts of the reserve. The salt marsh is particularly fine here, and its proximity to the road allows easy examination of the developing Glasswort swards and rich growths on more mature areas, of Sea Purslane (p. 118). On the left of the road there is the Bishop's Mill, one of at least ten tidal mills which once worked on the Lough.

The Dorn to Portaferry: Continue along the Abbacy Road, parting company with the Lough shore for about 3 km. A short way past the Abbacy itself, built within the shell of a very much older building, bear right, continuing roughly westwards towards the Lough shore. On reaching the shore, the route swings sharply south heading towards the entrance of the Lough. Continue along this road until finally reaching the waterfront at Portaferry.

Marlfield and Ballywhite shores: These shores recieve much of the force of south westerly gales blowing across from the Quoile, but they are fine areas to stop and admire the views across to the Quoile area with the ever present Mountains of Mourne beyond. To the south lies the beautiful Lecale shore of the Lough. Beyond, the Ordovician and Silurian bedrocks form the high ground of Lecale, continuing under the Narrows that link the Lough with the Irish Sea, and emerging as the 'Portaferry Mountain' rising to the left of the road at our viewpoint. In the waters off shore it is usually possible to see Black Guillemots fishing for Butterfish (p. 191).

Ballyhenry Island: (N.T.) This island, marked by a layby about 1.5km south of where the road rejoined the shore, is connected to the mainland, and marks the northern point of the turbulent Narrows. Similar to Horse Island in the wealth and variety of its vegetation and bird life, the outer point with its navigational beacon is swept by powerful tidal currents and an enormous variety of shore life can be found in a very small area, amongst the dense weed growth (p. 93) and tucked into the cracks of the tightly folded Silurian grits and shales. In the more sheltered corners lie soft muds and these have a particularly rich selection of burrowing species. Offshore lies the wreck of the Empire Tana (p. 63), rich in subtidal marine life, and usually perched on by a variety of fish eating sea birds. Divers who do not have the opportunity of boating there can have a very good substitute by swimming out to one of the two hulks that lie next to the south shore of the island. In summer the island is a good vantage point for watching terns dive for the shoals of Sandeels that are such a feature of the area.

The Pins: This is a site for land-based divers in Ballyhenry Bay. Find the first hole in the wall after the house with the green roof, and swim out about 25 m from the low water zone. The bottom is mixed mud and shingle, with occasional scallops. It is one of the few sites in the Lough for hermit crabs with their hitch-hiking sea anenomes (p. 102); Curled Octopus and juvenile Anglerfish may also be found.

The Walter Rock and Nugents Wood: These lie immediately to the north of Portaferry. The strength of the currents (p. 37) flowing past the Rock is readily apparent from the road, whilst the rock itself is a popular roost for Cormorants and Shags, and the nearby shores are regularly fished by Herons working between the rocks and in the pools. Nugent's Wood (N.T.) is an attractive area of mixed woodland of Ash, Beech, Elm, Horse Chestnut and the ubiquitous Sycamore. It is especially attractive in early summer with dense drifts of Bluebells and Foxgloves, with Bugle and Harts-tongue Ferns. Occasional clearings provide intriguing glimpses of the Narrows below, and beyond Castle Ward House and its woodlands (N.T.).

Portaferry: There is much to see in this attractive village, and it is only possible to refer to a few places. We first encounter the Ferry Terminal that provides access to Strangford on the opposite side. Above the terminal, on the left, is the Queen's University Marine Biology Station. Immediately past this, turn left past the Portaferry Hotel and go about a hundred metres up Castle Street. On the left is the 16th century Portaferry Castle (D.o.E.(H.M.B.)) and next to it the Northern Ireland Aquarium (A.B.C.). This ambitious and highly successful establishment provides a fine introduction to the marine life of the Lough and the neighbouring coasts.

Portaferry Windmill: Before boarding the ferry it is well worth taking a diversion and heading further south out of the village towards the entrance of the Lough. After leaving the Aquarium drive up to the Square and turn right onto Ferry Street, leading back to the shore. Almost immediately turn left up Steel Dickson Avenue and at the top turn right onto Windmill Hill. This road leads out of Portaferry up to the old windmill, where there is a layby. From here there is a marvellous view of the Narrows and its tidal currents, with the main part of the Lough to the north and the Angus Rock and the Irish Sea to the south. To the east the ringfort on Tara Hill is clearly visible and (on a clear day) the South Rock and the Isle of Man can be seen.

Portaferry Windmill to Ballyquintin Point:
From the windmill descend southwards, turning right at the bottom to return to the shore via Cook Street. Turn left at 'The Scotsman' to follow the road south along the shore.

Granagh Bay: About 2km from Portaferry the view opens out to provide a panorama of the Granagh Bay Nature Reserve (D.o.E.(C.& W.), N.T.). As with the Dorn, a complex of rock outcrops and sheltered muds provides a rich habitat for marine life, this time enhanced by powerful tidal currents. Seals, both Common and Grey (p. 77), haul out in large numbers on the outer rocks, and flocks of terns dive for fish in the fast flowing channels.

Barr Hall Bay: (N.T.) Where the road next meets the shore, we get a clear view of the rocks and dangerous reefs that mark the entrance of the Lough. Most notable of these is the Angus Rock, (the lighthouse is maintained by the Irish Lights Authority) and to the east, the white pillar of the Pladdy Lug. Most boats enter and leave the Lough between these markers. The bay is a maze of rock outcrops and shingle spits, separating enclosures of coarse sands - the latter an indication of the restless nature of the waters here. Only in the most sheltered pockets are there fine muds. The area is often rich in bird life, depending considerably on the weather, some conditions creating shelter that attracts large numbers of waders, as well as open sea birds like Gannets. The surrounding rocks are covered in small coastal plants clinging to pockets of soil in the crevices (p. 123).

The bay is the most southerly point on this side of the Lough that can be reached by car. There is a small layby at the head of the bay, and this is best place to park.

Ballyquintin Point: From the layby a track leads towards the point; unauthorised vehicles are NOT permitted to use this, as it is dangerous and there are no turning places! Having said that, the walk is well worth while, although the latter part of the track is private, so the route crosses on to rocky shore (N.T.). Ballyquintin Point is a National Nature Reserve (D.o.E.(C.& W.).

particularly on account of its rather strange plant life (p. 125). The shore life is rich and varied, with some of the most exposed rock outcrops in the Lough sheltering pockets of windswept salt marsh. From the Point, there is a fine view of its counterpart, Killard Point (p. 126), guarding the western side of the Lough's entrance. It is possible for more hardy souls to walk right round Ballyquintin to Port Kelly and the Tongue on the outer coast, but this can only be done via a very rocky shore.

After visiting the Point, return to Portaferry via the shore road to catch the ferry. (Departures from Portaferry are quarter to, and quarter past, each hour. Departures from Strangford are on the hour and half past the hour. There may be more frequent sailings during peak periods in summer.)

The Ferry Crossing: There are a number of aspects of the Lough's wildlife that are best seen from the ferry. In summer, before the ferry leaves, look out for jellyfish, usually Moon Jellyfish (p. 56), carried past, and under, the boat by the current. On the outer face of the quay there are limpets, barnacles, Dogwelks, all typical residents of an exposed rocky shore situation (p. 91), and at low tide the kelps are visible.

The route taken by the ferry usually depends on the state of the tide. On an ebb tide, or slack water, the direct passage between Portaferry and Strangford will be taken, but on a flooding tide the currents are running against the ferry. At these times the ferry heads south, close to the Portaferry shore, avoiding the strongest currents. In the vicinity of the Cooke St. quay it turns westwards, and cuts across the main flow, usually losing ground to the current before reaching the quieter waters off Swan Island in Strangford harbour. If the tide is ebbing during the journeys from Strangford to Portaferry, a similar procedure is adopted, the ferry working its way against the current close to the north side of Strangford, the Catherine Dock, and Church Point, before swinging out across the channel.

After departure, look out for the turbulence in the water, some 100m out from the shore (p. 37). Seals may be fishing in the area, though they will normally dive some way in advance of the ferry. Cormorants, Shags, and in summer, terns, all fish out in the channel. Between May and late July, as the ferry approaches Swan Island, it is possible (especially from the upper deck of the ferry) to see the nesting colonies of Sandwich and Common Terns along with Black-headed Gulls (p. 187). For people without boats this is the most easily seen of the Lough's many nesting islands. It is possible to spot young terns and Oystercatchers running along the shore. In winter the island is used by roosting Herons and Lapwings. After disembarking, a good viewing point may be reached by turning right and following the road round to a quay on the other side of the harbour.

Strangford to Killard Point: Before resuming the main circuit of the Lough it is worth diverting to see the western shore of the Narrows and Killard Point, which despite the apparent similarity of its situation to that of Ballyquintin, is very different in character and wildlife.

On driving off the ferry bear left, and a short way further, turn left at Duffy's shop, on to the A2 signed to Ardglass.

Isle O'Valla: About 1 km out of the village, on the right, is the derelict Isle O'Valla house, with its small cluster of trees. In the spring these form the basis of a small Heronry (p. 174)-the birds in the tree tops can be seen from the road, along with a large number of Rooks.

Cloghy Rocks Nature Reserve: (D.o.E.(C.& W.) This is on the left a short way south of Isle O'Valla, accessed by a small car park. It is one of the best places on the west shore of the Lough for land-based visitors to view seals (p. 77), and to do this it is best not to wander onto the shore as they will move further away and be more difficult to see. The shore life is rich, with good examples of rocky shore communities contrasting with sheltered muds holding species like Heart Urchins.

Tully Hill Layby: (D.D.C.) This site, misleadingly referred to as a 'spoil heap' on the Ordnance Survey Map, is one of the best places on the west side of the Narrows to view wintering birds. A car quietly stopped about an hour before high water will be ignored by flocks of Ringed Plover, Golden Plover, Redshank, Dunlin,

Turnstones, Teal, Shelduck, and Brent Geese. Occasional visits by Peregrines however, can create mayhem. Many of the fields near the shore in this area also attract large flocks of waders, particularly Lapwing and Curlew.

Kilclief Castle: (D.o.E.(H.M.B.)) One of the finest examples of a mediaeval Tower House (15th century) near the Lough. In addition to the interest of the Castle itself, there are fine views across to the Angus Rock and Ballyquintin Point. On occasions the interaction between currents, waves and wind create spectacular seas across the Bar in this area (p. 45). In squally weather, particularly in autumn and spring, there are often Gannets to be seen fishing off the Angus. The shore supports a wide variety of sand and mud dwelling species, including Sand Mason Worms, whose tubes can be seen protruding from the sediments at low water. The salt marsh immediately to the north of here is well worth examining. To continue south from here, bear left along the shore road, signed to Ballyhornan.

Kilclief Bay: This attractive beach is one of the few suitable areas of traditional bathing-type sand in the Lough, and it is popular in good weather. In winter it is used by Ringed Plover and Dunlin.

Millquarter Bay and Killard Point National Nature Reserve: (D.o.E.(C.& W.) This is the most southerly part of the Lough. They can be reached by a narrow lane on the left about 2km south of the previous junction; it is probably better to leave the car at the turning, and walk down to the shore and out to the point. Alternatively, continue for 1km and park at the north end of Ballyhornan Bay. There is a good view of Gun's Island to the south.

Killard Point (p. 126) has a wealth of marine life on its shores, including that on some of the most exposed rocks accessible without a boat. Sharply contorted bedrocks and rocky pools contrast with the gentle curves of sandy meadows and crumbling glacial debris. Wildflowers and butterflies greet the visitor in summer, with Skylarks hovering high above. In winter the area is rich in overwintering birds, including Purple Sandpipers, usually seen with Turnstones. Look for Grey Plovers out at the far point, and Divers in the waters beyond.

From Killard Point, return to Strangford, turning left onto the Downpatrick Road by Duffy's.

Strangford to the Quoile Bridge: Take the A25, heading southwest out of the village. Either head straight for the Quoile on the A25, or about 5km out from the village bear left (signed to Raholp). This route heads over the high ground on the north side of Lecale and up the Quoile Estuary. All routes arrive at the river Quoile, which we follow (signed to Killyleagh and Comber) and shortly after turn right, to cross the Quoile Bridge (signed Killyleagh and Comber).

Castle Ward Bay: (N.T. & R.S.P.B.) In winter Brent Geese, Wigeon, Gadwall and Mallard all dabble in the Bay which lies just behind Strangford Village. During low water it is often possible to see Greenshank (p. 154) hunting in the stream that runs under the Black Causeway. From the eastern part, closest to the village, it is possible to walk all the way round its shore, round much of Church Point, and back into the village along the Strangford Bay Path. In addition to the birds, look out for the outcrops of bedrock with their dense growths of weeds. Beneath the weeds is a rich variety of marine life, typical of sheltered rocky shores (p. 93). Both Kingfishers and Herons use this area. The marsh to the left of the road is also used by the birds. A few yards further, on the right of the road, there is access to birdwatching facilities further into the bay, provided by the Trust and R.S.P.B..

Castle Ward House: (N.T.) This fine eighteenth century mansion with its bizarre combination of Classical and Gothick architecture, along with exhibitions about life below stairs in the Victorian era, is open from April to October. The estate itself is open all year round. Details may be obtained from the N.T.

Strangford Lough Barn: (N.T.) This is the main centre and exhibition of the National Trust's Strangford Lough Wildlife Scheme. Sited near the farmyard at the lower entrance to the estate, it is reached by going into Castle Ward. There are videos, a small bookshop, and an exhibition explaining the wildlife of the Lough and the work of the Scheme. A touch table (very popular with children!) gives visitors the opportunity to examine specimens from the Lough at first-hand.

Audleystown Cairn and Audley's Castle: (D.o.E.(H.M.B.)) These can be reached by continuing along the A25 - the Audleystown Road is the first public road on the right after the turning to Castle Ward. Alternatively if visiting Castle Ward, they can be reached by leaving by the northern Farmyard entrance and following the road first north, then west. Approaching from the A25, the former of these two sites (both well signposted) is a Neolithic long cairn where the remains of 34 skeletons were found in 1952. There is a small car park available. There are fine views north up the Lough, including Jackdaw Is., one of the most important nesting islands. Audley's Castle is a 15th century tower house with a superb view south down the Narrows and across to Portaferry. The surrounding woods have an attractive path that leads round the Point.

Rejoining the A25, there is now a choice. The A25 leads straight to the River Quoile, with a chance of briefly diverting to the right (signed to the Quoile Yacht Club) to see the lower reaches of the Quoile Pondage. Alternatively, some 500m after the Audleystown Road junction, bear left to head up to the high ground of the Lecale and Slieve Patrick, reached after 2km. The route then continues to a fork 1km later (take the Mearne Road to the right) in Saul, and about 1.5km later rejoins the A25 at the Quoile.

Slieve Patrick: (Parish of Saul and Ballee) This is one of the highest hills of Silurian bedrock in the area and it has superb views of Strangford Lough with the drumlin islands (p. 22) off Killyleagh to the north and to the Mountains of Mourne in the south west. It is said to be the site where St. Patrick first preached and there is a large statue of him at the top of the hill. St. Patrick's Way (D.D.C.) is an attractive walk incorporating this and many other locations, and is a good way of seeing this countryside, with its outcrops of bedrock, banks of Gorse, and tiny fields. For those who love Peregrines and Ravens, Slievenagriddle, roughly 2 km walk to the south is another fine vantage point. The two species compete vigorously for nesting sites in the cliff, and aerial battles in the spring are quite common.

Quoile Pondage: (D.o.E.(C.& W.)) The story of the Pondage and its wildlife formed through the erection of a tidal barrage (p. 137) is explained in an exhibition at the Quoile Centre. Those traveling on the A25 will find it signposted to the right shortly before Downpatrick; those taking the Slieve Patrick route should turn right at the Quoile and head 500m back towards Strangford; for them it is then signed to the left. There is a good selection of paths, picnic areas and viewing points within the Nature Reserve, although there is a large section on the other side of the river where the wildlife has priority, and regular visitor access is not permitted.

From the Centre return to the A25, and head towards Downpatrick. After about 500m, turn right (signed to Killyleagh and Comber) and this follows the river for about 1km.

Jane's Shore and Quoile Bridge: There is a path along all of this section of the river, downstream managed by the D.o.E., and upstream, including Jane's Shore, by the National Trust. Much is wooded with Willows, Alder, Beech, Hazel and Oak; and there is a good chance of seeing Great-crested Grebes and Kingfishers. The attractive 17th Century bridge reached by turning right on to the A22 (signed to Killyleagh and Comber) carries us out of Lecale and north along the western shore of Strangford Lough.

Quoile Bridge to Killyleagh: Immediately after the Bridge bear right, continuing north on the A22 passing through the woodland of the Finnebrogue Estate. Its freshwater lake is linked with the Pondage, and thus was once connected to the Lough, and on the right of the road there is an old tidal watermill. Some 3km after the bridge there are further fine views of the Pondage, and from here in winter it is possible to see some of the large flocks of Wigeon, Mallard and Teal which come in to feed.

Gibb's Island: (N.T.) At the first true crossroads about 5 km after the bridge, turn right into Island Road, and continue to the bottom. From there a path leads out to Gibb's Island, with its distinctive landmark of Scot's Pines. The shores in this area are extremely sheltered. Because of the Sluice Gates of the Pondage nearby, the water is brackish and species tolerant of freshwater are much in evidence here. There are some quite well developed areas of salt marsh and salty meadows on the low spit leading out to the island. Teal are perhaps the most frequently seen wildfowl, whilst the most notable waders are Turnstone, Redshank, and Greenshank. Look out for Cormorants passing high overhead as they commute up the river to roosts on the lakes in Seaforde Demesne and Dundrum.

Delamont: (D.D.C., B.E.L.B., N.T.) Returning to the A22, the entrance to Delamont Estate is on the right within 1km. This is now a country park, and from the point of view of Strangford's wildlife, the most important part of it is the heronry in Kinnegar Wood (N.T.), down by the shore, with over thirty nests high in Scot's Pines (p. 174). The best way to see these is not to go into the wood, but to remain up on the slopes behind near the house, where one is almost looking down onto the wood.

From Delamont, return to the A22 and head north to Killyleagh.

Killyleagh to Whiterock: Continue through Killyleagh along the A22 (at the main crossroads in the town, look left up the street to see the fairy-tale Killyleagh Castle (P.)). A little over 2.5km out of the town turn right onto the Ringdufferin Road. This section of the route lies in the heart of the drumlin country on the west of the Lough, where the deposits of glacial debris are at their thickest (p. 22). Gone are the open rocky shores of the Narrows; a maze of winding narrow lanes takes us to an intricate coastline of sheltered interlocking bays and rounded headlands.

The first of these is Ringdufferin Bay, and immediately after the road leaves the Bay it bears sharply left, and winds through the tightly knit drumlins, intersperced with small damp hollows (p. 135). A little after 1km the road turns sharply left, and almost immediately right. Continue along this road which eventually changes its name to Ballymorran Road (the first turning on the right leads to the yachting centre of Ringhaddy Sound), and take the second on the right, Quarterland Road, which leads down to the inner reaches of Quarterland Bay.

Continue past the bay, and under 1km distance turn right, rejoining Ballymorran Road. This goes past Ballymorran Bay, and up to the top of a high drumlin which has a superb view of the whole complex of drumlin islands and pladdies that lie off the western shore (p. 128). In the middle of the Lough we see Bird Island, and beyond Kircubbin, with Greyabbey slightly to the north. In clear weather it may be possible to see the high ground of Lecale and southern Ards in the far distance.

Ringdufferin Bay: (P) 1km from the turning, this little bay beautifully illustrates the shelter to marine life in this part of the Lough, with luxuriant growths of weed on boulder shores alternating with banks and channels of soft mud. Look for Teal in the water channels, whilst further out there may be Greylag and Canada Geese. Both the shore and land are private, and the area is maintained as a bird sanctuary, so it is important not to disturb the bird life.

Quarterland and Ballymorran Bays: These are two of the most sheltered bays in the Lough. The smooth, regular slopes of these drumlins contrast with the notched examples of those on the eastern shore with their wave-eroded western faces and raised beaches (p. 156). At low tide, the surface of the mud is soft and glistening and cut deep by small streams; the sediments are too fine for larger animals like Lugworms, but there is an infinite wealth of microscopic life and organic matter. Wintering Golden Plovers and Lapwing use the grassland areas, whilst it is possible to find small flocks of Black-tailed Godwit (p. 154) along with Shelduck, leaving erratic tracks across the mud surface.

At the end of Ballymorran Road there is a 'T' junction. Turn right and head eastwards into Whiterock.

Whiterock to Ballydrain: A sharp left hand corner at the bottom of the hill brings us to the Whiterock shore (be careful; some people have been known to drive straight on!). This is the largest yachting centre on the Lough, and it is a popular base for many people to go exploring amongst the Lough's islands.

Continue northwards round the bay. About 1km further on there is a causeway to Sketrick Island with its 15th century towerhouse, and the headquarters of Cuan Sea Fisheries, famed for their locally grown oysters. Shortly after the road passes the South Rock Lightship, formerly anchored off Portavogie, now the elegant club house of Down Cruising Club. On the other side of this fast flowing channel (p. 39) lies Rainey Island (U.W.T.), with its old Oak Wood, one of the last examples of this habitat by the Lough shore. Beyond is Mahee Island that separates Ardmillan Bay from the rest of the Lough. From here the road gradually bends westwards, and after 1.5 km the route takes a right turn, followed almost immediately by another, and we arrive in Ardmillan Village.

Heading out of Ardmillan, bear right, and for the next 2 km there are intermittent views of Ardmillan Bay, and a small lane leading to the ruins of the late mediaeval Tullynakill Church (p. 14), where it is said that 'Daft Eddie', the smuggler's boy turned local hero is buried. Continue for a further 2 km into Ballydrain, turning right at the junction.

Whiterock Bay: A popular sailing centre, but in spite of all the recreational pressures, it is popular with wintering bird life. The best location is opposite a small salt marsh island north of the yacht club. Here Teal, Brent, Redshank, and Snipe may be seen, whilst further out, closer to Hen Island, there are Goldeneye and sometimes Slavonian Grebe.

Ardmillan Bay: This, by a wide margin, is the largest of the bays of the Lough. Most of its open mudflats are too soft to walk on. It has rich stocks of eelgrasses and *Enteromorpha* and is an important feeding ground for Brent Geese, Wigeon and both Whooper and Mute Swans. The area is important for waders also, with Bar-tailed Godwit and Curlew being seen in large numbers, and Black-tailed Godwit in the most sheltered corners. At a number of sites around the bay there are dense swards of the spikey looking *Spartina* grass (p. 120) which threatens to colonize much of the Bay. This alien is of little value to wildlife, although occasionally Snipe may be seen sheltering in it.

Ballydrain to Nendrum: Continue through the small cluster of houses and barns that constitutes the village and turn right on to Ringneill Road, signed to Nendrum. This road goes all the way out to the drumlin islands that encircle Ardmillan Bay, and for the unintiated it may be a confusing experience as the road winds across islands and causeways with the main Lough and the Bay seemingly encountered in all directions. These areas of mudflats, and lowlying pladdies are some of the richest areas for burrowing invertebrates in the Lough (p. 107), and in winter the bird populations are correspondingly important. There is a small carpark at Nendrum and from here we return to Ballydrain.

Ringneill Bay: The first area of shore and sea to be encountered is a small arm of Ardmillan Bay on the right of the road, extensively overgrown by *Spartina*. Look out for wildfowl feeding beyond this area.

Ringneill Quay: (N.T.) This small quay is on the left of the road, pointing out into the main part of the Lough. It is a good example of the number of small quays and slipways built in the last century for the shipping of farm products and coal, when landbased communications were poor. In winter the bay beyond has some of the most varied populations of overwintering birds.

Island Reagh: Immediately after the quay, the road crosses onto Island Reagh and soon cuts across onto the eastern side of the island. Sheltered bays make for excellent bird watching, but the best times are about two hours before and after high water. On the left of the road, about 500m south of the causeway leading to Cross Island there is a carpark and hide (N.T.) with facilities for the disabled. Look out for Pintail, Shoveler and Black-tailed Godwit on the rocks in the centre of the bay.

Nendrum Abbey: (D.o.E.(H.M.B.)) This is one of the most important early Christian sites in Ireland, founded by St. Mochaoi in the fifth century. It is believed to have been raided by Vikings in 976, and although it was used for a period after this, it never regained its former status. From the ruins, particularly the stump of the round tower, there are fine views of Ardmillan Bay, and its channel past Rainey Island connecting it to the Lough. From Nendrum, return to Ballydrain.

Ballydrain to Comber: Arriving in Ballydrain, turn right, heading north towards Comber, and the drumlin scenery begins to give way to more open country. In under 1.5km the route passes Castle Espie, after a further 1.5km there are good views of the Comber River estuary with the wide stretches of reclaimed land, and about 2km after that the road meets the A22. To return to Comber, turn right here and continue straight on, and those returning to Belfast should turn left some 500m further, continuing with the A22.

Castle Espie: (W.W.T.) This is an extensive wildfowl collection, bird sanctuary, and art gallery, with a restaurant and shop. The facilities are popular with both people and wild birds. There are hides for watching the birdlife on the lakes, and these are likely locations for occasional rarities, perhaps the most notable being the visit by an American Lesser Yellowlegs several years ago. The ponds are the former quarries for the excavation of Carboniferous limestone and boulder clay (p. 14).

Montgomery Hide: (N.T.) This is a small stone-built hide reached by taking the lane on the right immediately after Castle Espie. From here there are good views of both wildfowl and waders, but it is best to be there two hours before or after high water, or the birds may be far out on the mudflats. This is an immensely rich feeding ground; extensive eelgrass and *Enteromorpha* swards stretch into the distance and the mud is thickly populated with Lugworms, small shrimps and snails. Near the hide are some good examples of the Lough's salt marshes (p. 115), with Sea Asters, Lax-flowered Sea Lavender, and Sea Pinks all coming into bloom at various times of the year.

Comber River Estuary: A lowlying area of salt marshes, with the Comber River meandering between banks of soft mud. Where small streams meander down to the River they are fringed by thickets of reeds, and behind these, the open fields of reclaimed farmland. At high tide the marshes may be completely inundated, and the River only visible as an area of slight turbulence where it interacts with the incoming Lough waters. The salt marshes are rich in wildflowers and these and the mudflats are important for bird life. Early in winter the area is used extensively by Wigeon and Brent Geese, whilst throughout the winter large flocks of waders feed on the invertebrates. At high tide look out for large flocks of Golden Plover and Lapwing flying high over the estuary and the adjacent land (p. 162).

A TOUR BY SEA

General Points; This tour is based on a journey from the Narrows in the south of the Lough, northwards along the western side, and across into the centre of the Lough, then southwards back to the entrance. The information presented here is only a summary and the Admiralty Chart 'Northern Ireland - East Coast, Strangford Lough' (Sheet No. 2156) is essential, whilst the larger scale Chart 'Strangford Narrows' (Sheet No. 2159) is extremely helpful in this difficult part of the Lough. Visitors should carefully follow these, and the Irish Cruising Club's Sailing directions for the North and East Coast of Ireland on their first visit, and should only enter on a rising tide. If in doubt seek local advice.

A large number of islands are privately owned, and permission should be sought from the owners before landing. Many islands are owned by the National Trust; some of these are important islands for bird life and during the nesting season, signs are erected asking people not to land so that the birds may nest undisturbed. However, there are others (some indicated below) where visitors are encouraged to land and explore. Details of these may be obtained from the Trust.

The Narrows: The connection between Strangford Lough and the Irish Sea. The centre of the channel is generally deep, but there are numerous rocks and shallows along the edge and about the Angus Rock.

Tidal currents exceed 8 knots in places and may be very turbulent (p. 36). Towards the northern end are the villages of Portaferry and Strangford, and further to the north the opening into the main body of the Lough.

The channel cuts straight through the underlying bedrocks of the Lough, which can be seen emerging as the high ground of the Lecale and southern Ards Peninsula. From the boat, the outcrops of tightly folded bedrock can be seen along the shore of much of the Narrows.

Angus Rock: A long outcrop of tightly folded bedrock at the entrance of the Lough, marked by a lighthouse at the northern end, and a pillar on the Garter Rock to the south. Extremely rich in shore life, not least because many types of subtidal life manage to live in the intertidal zone here. The seals are easily seen from a distance with up to a hundred Common Seals occuring frequently, regularly accompanied by smaller numbers of Grey Seals.

Cloghy Rocks: (D.o.E.(C.& W.)) (p. 75) These lie off the western shore of the Narrows, about 2 km south of Strangford. It is hazardous for boats to venture among their rocky channels, but the seals may be seen from off the beacon at their outer limit.

Granagh Bay and the Gowland Rocks: (D.o.E.(C.& W.), N.T.) (p. 75) These lie opposite the Cloghy Rocks and are also hazardous for boats. However, the seals are if anything even more visible, since they can often be seen lying below the Gowland rocks beacon.

Routen Wheel: Lying immediately to the north of Granagh Bay, this is a whirlpool of some considerable force, particularly during spring tides. It is created by the turbulence caused by irregularities on the bottom some 15m below, although there is a shoal where the bottom rises to a mere 4m depth.

Swan Island: (p. 187) A nesting island populated by terns, gulls and Oystercatchers, lying in the centre of Strangford Harbour amongst the moorings. As with many other important nesting islands, the National Trust and the D.o.E erect signs during the nesting season requesting people not to land, so as to allow the birds safe and undisturbed nesting. However, excellent views may be had of the birds from a boat.

Castle Ward Estate: This can be seen at the head of Castle Ward Bay on the western shore of the Narrows to the north of Strangford. The National Trust welcomes visitors by boat, and the Strangford Lough Wildlife Scheme has its centre in the Strangford Barn in the estate, with its exhibition and videos about the Lough's wildlife.

Ballyhenry Wreck: (p. 63) This lies off the eastern shore of the Narrows to the north of Portaferry and to the south of Ballyhenry Island. It is one of the most popular dives in the Lough, but the wreck is collapsing dangerously so particular care is needed if venturing inside. However, in spite of the currents it can make for an interesting dive even for the less adventurous. Maximum depth is about 15m, best time for diving from three hours after high water. After low water there is a particularly strong current around the stern of the wreck.

Ballyhenry Island: For those wishing a deeper dive in the area, there is a steep bedrock and boulder slope to the south west of the Island's beacon, that descends eventually to 59m. The area is a luxuriant mass of sponge growths, Plumose Anenomes, and hydroids. Recommended time for diving is about half an hour before low water - at other times there are powerful currents.

To the north of Ballyhenry Island the Narrows quickly opens out into the main body of the Lough.

Along the western shores: From the northern end of the Narrows head approximately northwest towards Dunnyneill Island some 2.5km away. Avoid the Long Rock off its southwestern shore. From here the route will pass Taggart Island about 1km from Dunnyneill, and head northwards past Black Rock and the Ringdufferin Estate, through Ringhaddy Sound and thence to Darragh Island.

Dunnyneill Island: (P.) This may be approached in either the outward or inward bound sections of the journey. There are further details in the latter.

Taggart Island: (N.T.) This is one of the largest islands in the Lough and used to contain two small farms. Visitors are welcome here and there are good anchorages off the eastern shore, and the north west corner of the island depending on the weather, although care on a falling tide is recommended. Old farm buildings give a good indication of life on the islands, and indeed they were used by Little Bird Films to make 'December Bride', a story about County Down folk at the turn of the century. Thick hedges full of bird life, relatively unspoiled meadows full of wildflowers, and small marshes bright with Yellow Flag Iris and Orchids make this a lovely island to visit, whilst in high summer it is full of butterflies including large numbers of Common Blues and Small Coppers.

Black Rock: (P.) A tiny island lying about 1 km northeast of the north end of Taggart Is. Its northern end is surrounded by shallows and pladdies (The Brown Rocks) and there are often Common Seals hauled out on these.. The island is managed as part of the Ringdufferin Refuge area, and landing is not permitted. However, its thriving Cormorant colony (p. 175) can easily be seen by boat from both the east and west sides were the water is sufficiently deep to get close without landing or disturbing the birds. A colony of Common Terns has also become established.

From Black Rock, head northwards to Ringhaddy Sound, keeping clear of the rocky shores extending out from Ringdufferin's Castle Island. Pass through Ringhaddy Sound; there are moderately strong tidal currents running through here, and the whole area, besides being a major anchorage for cruising yachts, is a sort of M1 for the Lough's boating traffic, so care is needed. To the northern end on the western shore, look for the ruins of Ringhaddy Castle, a rare example of an Elizabethan castle in the Province.

Darragh Island: (N.T.) Lying immediately to the north of Ringhaddy Sound this island, though not particularly important for nesting birds, has some of the richest terrestrial wildlife of any of the Lough's islands (p. 130). There is a sheltered anchorage in the bay on the northwest side of the island, but take care on a falling tide.

The western shore to the central channel: From Darragh Island head east through the scatter of islands off Whiterock. There are numerous ways of doing this; those wishing to make a longer run might find the passage east of Roe Is. and Inisharoan Is attractive, and this more northern entry into the central channel would allow for views of Bird Is. with its large Cormorant colony. Those with less time might use the channel between Green Is. and Great Minnis, to the north of the Sand Rock Pladdy and the Hadd Rock. In any case it is worth keeping an eye out for Porpoises which are sometimes reported in the area, and in winter the area is used extensively by Brent Geese, Wigeon, and Teal, in addition to small flocks of waders.

South by the central channel: Head south, leaving the islands off Whiterock about 1 km to the west, and a similar distance to the maze of pladdies to the south of Bird Is. to the east. Those wishing to divert to Kircubbin may find their best route is, with care, via Bird Island Passage. The main route south passes the Long Sheelah, and from there it is more open, though there are a number of pladdies like the Abbey Rock to trap the unwary. Ballyhenry Island and the entrance back into the Narrows lie about 6km south of the Long Sheelah.

Slave Rock: One of the most southern of the complex of pladdies south of Bird Island, this is an excellent place for diving to explore the rich Horse Mussel community. Although the community has been badly damaged by trawling in the central channel, this obstacle has given some protection to the marine life nearby. The mussels themselves are overgrown with a wealth of other forms of life (P. 66)

From the Slave Rock head south, past the Long Sheelah. Look out for Cormorants and Shags roosting on this long spit. There may also be large numbers of Common Guillemots and Razorbills in the area, and in winter Great Northern Divers as well.

Limestone Rock: Another good diving location about 200°, and 3 km, from off the Long Sheelah. Low water slack is the best time, and divers should swim north east from the rock to descend over a mass of rocks and ledges, heavily overgrown with kelps near the surface, and below, 'hedgehog stones', overgrown with pink calcareous algae (p. 92), and numerous crustaceans.

Head south west to Dunnyneill Island, about 1km away.

Dunnyneill Island: (P.) This small island is in two sections, connected by a shingle spit. The lower, eastward section is often used by nesting birds, including gulls, wildfowl, waders and terns. The larger, high part of the island with its small wood is used by nesting wildfowl and gulls in the early part of the season. Later in the year it is popular as a picnic spot, particularly as the anchorage is safe and deep on the north side of the island. In the middle of the wood there is an earth bank and ditch, possibly neolithic, although mesolithic flints have also been found there. The shores are composed of a rich variety of materials - look for granites, limestones, Triassic sandstones, and fossil sea lily stalks, all of which have been carried there by the glaciers.

From Dunnyneill return to Ballyhenry Is. and the Narrows, about 2.5 km to the south east. On the way, look out for Black Guillemots fishing off Marlfield, and in late summer shoals of Sandeels and Herring fry may attract large flocks of terns and other sea birds diving for the small fish.

Some Books Recommended for Further Reading

General Books on the Coast and Countryside

Sea Life of Britain and Ireland: ed. Elizabeth Wood. 240pp. Marine Conservation Society; Immel Publishing, 1988.
The National Trust Guide to the Coast: Tony Soper. 224pp. Webb and Bower, 1984.
An Introduction to Coastal Ecology: P.J.S.Boaden & R.Seed. 218pp. Blackie, 1985.
The Sea shore: C.M.Yonge. 350pp. New Naturalist Series.
The Sea Coast: J.A.Steers. New Naturalist Series.
The Open Sea: A.Hardy. New Naturalist Series.
The Good Beach Guide: A.Scott. 189pp. Marine Conservation Society, Ebury Press, 1989.
The Ulster Countryside: C.D.Deane. 170pp. Century Books, 1983.
Discover Northern Ireland: E.Sandford. 218pp. Northern Ireland Tourist Board, 1976.
Sailing Directions for the North and East Coast of Ireland: Published by Irish Cruising Club Publications Ltd., on behalf of the Irish Cruising Club. approx 150pp, Updated annually.
Book of the Irish Countryside: by various authors. 288pp. Blackstaff Press/Town House Books, 1987.
Walking the Ulster Way: A.Warner. 184pp. Appletree Press, 1989.
Birds of the Grey Wind: E.A.Armstrong. 174pp. Lindsay Drummond, 1946.
The Quoile Quest: Department of the Environment (N.I.). 24pp. 1987
Scrabo Country Park: Dept. of Environment for Northern Ireland. 20pp. 1986.
National Nature Reserve Leaflets: Countryside and Wildlife Branch, Dept. of Environment for Northern Ireland, Belfast.
Diocese of Down and Connor, Vol. I: J.O'Laverty, 1878. 448pp. Facsimile by Davidson Books, Ballynahinch 1980.
A History of the County of Down: A.Knox, 1875. 724pp. Facsimile by Davidson Books, Ballynahinch, 1982.
The Antient and Present State of the County of Down: W.Harris, 1744. 185pp. Facsimile by Davidson Books, Ballynahinch 1977.
Our Countryside Our Concern: K.Milton, 96pp. A Report for Northern Ireland Environment Link, 1990.
In Search of Neptune. C. Pye-Smith. 252 pp. National Trust, 1990.

Identification Books

The Wild Flower Key: F.Rose. 480pp. Warne, 1981.
Guide to Inshore Marine Life: D.Erwin & B.Picton. 120pp. Marine Conservation Society; Immel Publishing, 1987.
British Bivalve Seashells: N.Tebble. 212pp. British Museum; H.M.S.O., 1966.
British Shells: N.McMillan. 196pp. Warne, 1968
Collins Pocket guide to the Sea Shore: J.Barrett & C.M.Yonge. 272pp. Collins, 1958.
Collins Guide to Sea Fishes of Britain and N.W.Europe: B.Muus and P.Dahlstrom. Collins.
Tracks and Signs of the birds of Britain and Europe: R.Brown, J.Ferguson, M.Lawrence and D.Lees. 232pp. Christopher Helm, 1987.
Gulls - a guide to identification: P.J.Grant. 280pp. T.& A.B. Poyser, 1982.
A field Guide to the Birds of Britain and Europe: R.Peterson, G.Mountfort and P.A.D.Hollom. 344pp. Collins, 1965.
British Whales, Dolphins & Porpoises: F.C.Fraser. 34pp. British Museum (Natural History), 1976.

Wildlife and Natural History Accounts

An Irish Beast Book: J.Fairley. 334pp. Blackstaff Press, 1984.
British Seals: H.R.Hewer. 256pp. New Naturalist Series.
Estuary Birds of Britain and Ireland: ed. A.J.Prater. 440pp. T. & A.B.Poyser, 1981.
The Seabirds of Britain and Ireland: S.Cramp. W.R.P.Bourne & D.Saunders. 287pp. Collins, 1974.
Flowers of the Coast: I.Hepburn. New Naturalist Series.
Waders: W.Hale. New Naturalist Series.
Seabirds: J.Fisher and R.M.Lockley. New Naturalist Series.
The Natural History of Whales and Dolphins: P.G.Evans. 343pp. Christopher Helm, 1987.
Birds Around Belfast: Belfast R.S.P.B. Members' Group. 66pp. 1981.
British Regional Geology - Northern Ireland: H.E.Wilson, Geological Survey of Northern Ireland. 115pp. H.M.S.O., 1972.
The Inshore Marine Life of Northern Ireland: D.G. Erwin, B.E. Picton, D.W. Connor, C.M. Howson, P. Gilleece, M.J. Bogues. H.M.S.O., 1990

Index

Accipiter nisus 149
Acer pseudoplatanus 128
Acrocephalus schoenobaenus 120
Actinia equina 96
Actinothoe sphyrodeta 58, 65
Adamsia palliata 102
Aequipecten opercularis 65
Aesculus hippocastanum 128
Agropyron repens 113
Alaria esculenta 92
Alauda arvensis 127
Alca torda 75
Alcedo atthis 88
Alcyonium digitatum 58, 60
Alder 30, 136, 138
Alexanders 128, 169, 170
Alnus glutinosa 30
Alopex lagopus 24
Ammodytes tobianus 63
Amphipods 95, 114
Amphitrite johnstoni 102
Amphiura filiformis 70
Anacamptis pyramidalis 126
Anas clypeata 146, 151
Anas crecca 137, 146
Anas penelope 107, 138, 146
Anas platyrhynchos 107, 137, 146, 171
Anas strepera 146, 171, 172
Anglerfish 68, 69, 71
Anguilla anguilla 71
Annelid worms 103
Annual Sea Blite 113, 118
Anomia ephippium 102
Anser albifrons 146, 152
Anser anser 137, 146, 171
Anser leucopsis 146, 169, 171
Antedon bifida 8, 101
Anthocharis cardamines 131
Anthus petrosus 127
Anthyllis vulneraria 124
Aphantopus hyperantus 131
Aporrhais pespelicani 70
Arctic Fox 24, 28

Arctic Skua 189
Ardea cinerea 87, 171
Arenaria interpres 139, 146
Arenicola marina 88, 107
Armeria maritima 114, 118
Arrow Worms 57
Ascidiella aspersa 69
Ascophyllum nodosum 95
Ascophyllum nodosum mackii 95
Ash 30, 130, 133, 135, 138
Asio flammeus 119
Asio otus 132
Astarte borealis 28
Aster tripolium 113
Asterias rubens 66
Asterina gibbosa 99
Atriplex prostrata 115
Aurelia aurita 56
Auroch 21
Aythya ferina 138, 146
Aythya fuligula 146, 151, 170, 171
Badger 88, 126, 134, 169
Balanus balanoides 92
Ballan Wrasse 65
Baltic Tellin 107, 148, 159
Bar-tailed Godwit 139, 143, 145-147, 153, 161, 162, 178
Barnacle Goose 146, 169, 171, 173
Barnacles 61, 92, 101
Basking Shark 72
Beadlet Anenome 96
Beech 132
Bell Heather 124, 131
Betula nana 28
Birch 28
Bird's Foot Trefoil 124
Biting Stonecrop 123
Blackfish, see Pilot Whales
Black Guillemots 75, 171, 191, 192
Black-headed Gull 149, 171, 178, 179, 188
Black Knapweed 124
Black Redstart 128

Black Serpent-star 65
Black-tailed Godwit 146, 149, 154, 162
Blackthorn 125, 130, 135, 168
Bladder Wrack 95, 98, 101
Blennius pholis 5, 96
Blochen, see Coalfish
Bluebell 132
Blue-rayed Limpet 98
Bog Bean 136
Bog Cotton 136
Boring Sponge 63
Bos primigenius 21
Bottle-nosed Dolphin 79
Brachiopods 8
Bramble 133, 135, 170, 173
Branta bernicla hrota 2, 107, 146
Branta canadensis 146, 152, 171
Breadcrumb Sponge 99
Bristle-tail 88
Broad-clawed Squat Lobster 100
Brown Bear 28
Brown Rat 88, 130, 169, 170
Brown-tipped Banded Snail 127
Bryozoans 98
Buccinum undatum 61
Bucephala clangula 146, 151
Burnet Rose 125
Buteo buteo 132
Butterfish 65, 83, 96, 192
Buzzard 132
Caa'ing Whale, see Pilot Whale
Calidris alpina 119, 146
Calidris canutus 140, 146
Calidris maritima 146, 154
Callimorpha jacobaeae 124
Caloplaca 92, 122
Canada Goose 146, 152, 171–173, 192
Cancer pagurus 68
Canis lupus 28
Carcinus maenas 96
Cardamine pratensis 131
Carduelis cannabina 114
Carduelis flavirostris 119
Carpet-shell 28
Carrion Crow 155
Celandine 130
Centaurea nigra 124
Cephalopods 11, 14
Cepphus grylle 75, 171
Cerastoderma edule 88, 107
Cerianthus lloydii 70
Carrageen 98
Cervus elephas 88

Cetorhinus maximum 72
Channelled Wrack 85, 94, 114
Charadrius hiaticula 143, 146, 171
Chelon labrosus 72
Chestnut 128
Chitons 102
Chlamys varia 67, 68
Chondrus crispus 98
Chthalamus stellatus 92
Ciliates 103
Clam, see Great Scallop
Cinnabar Moth 124
Clava squamata 98
Clavelina lepadiformis 63
Clione cellata 63
Clouded Yellow 127
Clover 124
Clupea harengus 71
Coalfish 72
Cochlearia officinalis 114
Cockle 88, 107, 109, 141, 158, 159, 165, 193
Cod 72
Codium tormentosum 101
Coeloglossum viride 126
Coelopa frigida 114
Coenonympha pamphilus 133
Coenonympha tullia 30
Coley, see Coalfish
Colias crocea 127
Common Blenny 5, 96
Common Blue 124, 127, 131
Common Brittle Star 65
Common Guillemot 75, 193
Common Gull 171, 179, 180
Common Hermit Crab 68, 102
Common Reed 119
Common Salt-marsh Grass 117-119, 169
Common Scurvey Grass 114, 115, 128, 169
Common Seal 2, 74-78
Common Sea Urchin 63, 100
Common Shore Crab 96, 195
Common Star 66, 100
Common Twayblade 126
Conger conger 65
Conger Eel 65
Coot 138
Copepods 56, 103
Corallina officinalis 92
Corals
Cord Grass 120, 121
Cormorants 37, 74, 88, 169, 171, 175-177, 179, 190, 193

Index

Corncrake 134
Corophium volutator 108
Corvus corone corone 155
Corvus corone cornix 155
Corvus frugilegus 155
Corvus monedula 127
Corylus arellana 30
Couch Grass, see Scutch Grass
Crab's eye Lichen 123
Crassostrea gigas 200
Crategus monogyra 133
Crawfish 58
Crex crex 134
Crocuta crocuta 28
Crossaster papposus 66
Cuckoo 177
Cuckoo Wrasse 62, 65
Cuculus canorus 177
Curled Dock 113
Curled Octopus 61, 66, 68, 100
Curlew 140, 146, 148, 149, 152, 153, 162, 178, 194
Cushion Star 99
Cyanea lamarkii 55, 56
Cyanea capillata 56
Cygnus cygnus 142, 146
Cygnus olor 137, 146, 171
Cynoglossum officinale 126
Dabchick 88, 155
Dabberlocks 92
Dabs 71, 109
Dactylorhiza fuchsii 126
Dactylorhiza incarnata 126
Dead-man's Fingers 58, 60, 63
Delessaria sanguinea 98
Dendrodoa grossularia 92, 100
Devil's-Bit Scabious 124, 131, 136
Diatoms 54, 55, 104
Dinoflagellates 55
Dog Cockle 65
Dog Whelk 92, 96
Dublin Bay Prawn 69, 70, 200
Dulse 97
Dunlin 119, 139, 140, 143, 146, 147, 149 152, 159, 160, 162, 163, 178, 194
Echinocardium caudatum 108
Echinus esculentus 63
Edible Crab 68, 92
Eel 71, 77, 100, 138, 176
Eel-grass 105, 117, 141, 157, 165
Eider 170, 171, 190
Elder 128
Eledone cirrhosa 61
Elephant's Ear Sponge 63

Elm 30
Elminius modestus 101
Emberiza citrinella 114
Emberiza schoeniclus 135
Emperor Moth 126
English Stonecrop 112, 123, 128
Enteromorpha spp. 107, 117, 121, 150, 158
Erebia epiphron 30
Erica cinerea 124
Eriophorum angustifolium 136
Euphydryas aurinia 131
Euonymus europaeus 130
Fagus sylvatica 132
Falco columarius 120
Falco peregrinus 149
Fan worms
Father Lashers 63, 71, 177
Feather Stars 8, 101
Festuca rubra 115, 118
Fieldfare 134, 149
Field Vole 29
Flabelligerina affinis 68
Flounder 71, 109
Football-Jersey Worm 68
Fox 88, 132, 134, 150, 169, 170, 190
Fratercula arctica 75
Fraxinus excelsior 30
Fucus serratus 89, 96
Fucus spiralis 95, 114
Fucus vesiculosis 95
Fulica atra 138
Fulmar 127
Fulmarus glacialis 127
Furbelows 96
Gadus morhua 72
Gadwall 146, 171, 172
Galathea squamifera 99
Galathea strigosa 68
Galeorhinus galeus 72
Gallinago gallinago 121, 146
Gannet 75
Gastrotrichs 103
Gavia stellata 75
Gavia immer 75
Geranium robertianum 125
Gibbula cineraria 102
Gigartina stellata 98
Gilpen, see Coalfish
Glashan, see Coalfish
Glasswort 116-120
Glaux maritima 118
Globicephala melaena 79
Glycymeris glycymeris 65
Goldeneye 146, 151, 166

Golden Plover 144, 145, 146, 154, 161, 162, 178
Gorse, 135, 136
  Common/European 124, 128
  Western 124
Gravel Sea Cucumber 65
Great Crested Grebe 138
Greater Black-backed Gull 171, 179, 190
Great Northern Diver 75
Great Scallop 65
Great Spider Crab 68
Greenland White-fronted Goose 146, 152
Greylag Goose 137, 138, 146, 152, 171, 172
Greenshank 146, 154, 178, 196
Greyling 127
Greylord, see Coalfish
Grey Mullet 72, 77, 119, 138
Grey Plover 143, 145, 146, 162
Grey Seal 2, 75-78
Grey Top Shell 102
Grey Wagtail 114
Haematopus ostralegus 140, 146, 171
Halberd-leaved Orache 115
Halichoerus grypus 2, 75-78
Halichondria panicea 99
Halimione portulacoides 118
Hare, Irish 125, 130
Hawthorn 133, 135
Hazel 30
Heart Urchin 108
Heather
Helix nemoralis 127
Herb Robert 125
Heron 87, 155, 166, 171, 174
Herring 71, 72, 169, 184
Herring-hog, see Pilot Whale
Herring Gull 128, 166, 171, 176, 179, 190
Himanthalia elongata 96
Hipparchia semele 127
Hippophae 25
Hirundo rustica 177
Holly 133
Honkenya peploides 114
Hooded Crow 155, 169, 189
Horse Mussel 29, 66-69, 195, 199
Houndstongue 126
Hyacinthoides non-scriptus 132
Hyas areneus 68
Hydrobates pelagicus 75
Hydrobia ulvae 107
Hydroids 63, 98, 103
Hyena 28
Hypericum elodes 136

Ilex aquifolium 133
Inarchus dorsettensis 68
Iris pseudacorus 112
Irish Deer 28
Jackdaw 127, 155
Juncus gerardii 118
Juncus maritimus 118
Kelp 92
Kelp Flies 114
Kidney Vetch 124
Killer Whale 79
Kingfisher 88, 119, 155
Kittiwake 75, 190
Knot 140-143, 146, 147, 149, 152, 159, 161, 162
Knotted Wrack 95, 98, 101
Labrus bergylta 65
Labrus mixtus 62, 65
Ladies Smock 131
Laminaria digitata 92
Laminaria hypoborea 92, 96
Lanice conchilega 102
Lapwing 146, 147, 154, 162, 171, 180, 182, 194
Large Heath 30
Larus argentatus 128, 171, 190
Larus canus 171, 179, 180
Larus fuscus 171, 179
Larus marinus 171, 179
Larus ridibundibus 149, 171, 178, 188, 190
Lasiommata megera 131
Laver 97
Laver-spire shell 107, 151
Lax-flowered Sea Lavender 112, 114, 118, 119
Lecanora atra 123
Lemmings 28
Lemmus lemmus 28
Lepadogaster bimaculatus 96
Lepidonotus squamatus 68
Leptasterias mulleri 29, 95, 98
Lepus timidus hibernicus 125
Lesser Sea Spurrey 115
Lichens
Lichina sp. 92
Light-bulb Sea Squirt 63
Ligia oceanica 87
Limander limander 71
Limpet 85, 92, 102
Limonium humile 112, 118
Limosa lapponica 139, 146
Limosa limosa 146, 149, 154
Linnet 114
Liocarcinus puber 63

Index

Lion's Mane 55
Lipura maritima 88
Listera ovata 126
Lithothamnion 92
Littorina littorea 96
Littorina obtusata 96
Littorina saxatilis 95
Lobster 77, 100, 200
Long-eared Owl 132
Long-legged Spider Crab 68
Lophius piscatorius 68
Lotus corniculatus 124
Lugworm 88, 106, 107, 109, 148, 153, 159
Lutra lutra 81
Lycaeana phlaeas 131
Mackerel 72
Macoma balthica 107
Macropodia rostrata 68
Mallard 107, 137, 146, 150-152, 156, 160, 165, 166, 170, 171, 173, 192
Mammoth 28
Mammuthus primigenius 28
Maniola jurtina 131
Manx Shearwater 75, 190
Marthasterias glacialis 61
Marsh Cinquefoil 136
Marsh Fritillary 131
Marsh Samphire, see Glasswort
Marsh St John's Wort 136
Matricaria maritima 115
Meadow Brown 131
Megaceros giganteus 28
Meles meles 88
Menyanthes trifoliata 136
Merlin 119
Metridium senile 60
Microstus agrestis 29
Modiolus modiolus 29, 66-69
Mole 29
Moon Jelly 56
Motacilla alba 114
Motacilla cinerea 114
Moss 30, 136
Mud Snail 107
Mustela erminea 125
Mya arenaria 107
Mycale spp. 58
Myoxycephalus scorpius 63
Mytilus edulis 88
Myxilla incrustans 99
Nematodes 103
Nephrops norvegicus 69, 70
Neopentadactyla mixta 65

Nereis virens 103
Nettle 128, 169, 182
Noctiluca scintillans 56
Norway Lobster, see Dublin Bay Prawn
Nucella lapillus 92
Numenius arquata 140, 146
Nyctea scandiaca 24
Oak 25, 30, 132, 138
Oarweed, see Kelp
Ochrolechia parella 123
Octopus Jelly 56
Oenanthe oenanthe 127
Onus mustelus 100
Ophiocomina nigra 65
Ophiothrix fragilis 65
Ophrys apifera 126
Orange Tip 131
Orchids,
  Bee 126, 129
  Common Spotted 126, 131
  Early Marsh 126
  Early Purple 131
  Frog 126
  Green Winged 126
  Pyramidal 126
Orchis mascula 131
Orchis morio 126
Orcinus orca 79
Orthoceras 14
Oryctolagus cuniculus 126
Otter 81, 88, 100, 119, 136, 150
Oystercatcher 140, 143, 144, 146, 152, 153, 154, 158-162, 165, 171, 180-182, 191, 193
Pachymatisma johnstonia 63
Pagurus bernhardus 68
Pagurus cuanensis 68
Painted Lady 127
Pale-bellied Brent Geese 2, 105, 107, 140 142, 145, 146, 148, 150, 156-162, 164, 166, 178, 194
Palinurus elephas 58
Palmeria palmata 97
Pararge aegeria 131
Passer montanus 134
Patella vulgata 85
Patina pelucida 98
Pea Crab 195
Pecten groenlandica 28
Pecten maximus 65
Pelicans-foot Shell 70
Pelvetia canaliculata 85, 92
Peregrine 149, 163, 169, 189

Perwinkle,
   Rough 95
   Smooth 96
   Edible 96
Peter Nine-eyes, see Butterfish
Petrobius maritimus 88
Phalacrocorax carbo 27, 74, 171
Phalacrocorax aristotelis 74
Phasianus colchicus 88
Pheasant 88
Phoca vitulina 2, 75-78
Phocoena phocoena 2, 78, 79
Phoenicurus ochruros 128
Pholis gunnellus 65
Phragmites australis 119
Phylloscopus trochilus 120
Pied Wagtail 114
Pigmy Shrew 88, 114
Pilot Whale 79
Pine 25, 30
Pinnotheres pisum 195
Pintail 146, 150, 160, 162, 165
Pinus spp. 25, 132
Plaice 71, 109
Plankton 14, 54
   Phytopl. 54
   Zoopl. 54
Plantago maritima 114, 118
Platichthys flesus 71
Plectropherax nivalis 119
Pleurobrachia pileus 57
Pleuronectes platessa 71
Plumose Anenome 60, 63
Pluvialis apricaria 145, 146
Pluvialis squatarola 143, 146
Pochard 138, 146, 151
Podiceps auritus 74
Podiceps cristatus 138
Polar Bears 24
Pollachius virens 72
Polyommatus icarus 124
Porcellana longicornis 68
Porcellana platycheles 100
Porphyra umbilicalis 97
Porpoise 2, 78, 79
Potentilla anserina 130
Potentilla palustris 136
Prostrate Orache 115, 118, 128, 178, 188, 190
Prunus spinosa 125
Psammechinus miliaris 99
Puccinellia maritima 117
Puffin 75
Puffinus puffinus 75

Purple Sandpiper 146, 154
Queen Scallop 65, 66, 199
Quercus petraea 25
Rabbit 126, 128, 130
Ragworm 103, 148
Raja batis 72
Ramelina siliquosa 123
Rangifer tarandus 28
Ranunculus ficaria 130
Rattus norvegicus 88
Razorbill 75, 193
Red-breasted Merganser 74, 88, 146, 165, 166, 170, 171
Red Deer 88
Red Fescue 115, 118
Redshank 143, 146-149, 161, 171, 180-183
Red-throated Diver 75
Redwing 134, 149
Reed Bunting 135
Reindeer 28
Rhizostoma octopus 56
Rhododendron 25
Rhododendron ponticum 25
Rhodymenia spp. 98
Ringed Plover 143, 146, 147, 154, 159, 162, 165, 171, 180-183, 191
Ringlet 131
Riparia riparia 127
Rissa tridactyla 75
Rockling 100
Rock Pipit 127, 155
Rook 155
Rosa pimpinellifolia 125
Rubus fruticosus 133
Rumex crispus 113
Ruppia maritima 106
Sacchoriza polyschides 96
Saddle Oyster 102
Sagartia 92
Sagitta elegans 57
Saithe, see Coalfish
Salicornia spp. 116
Salix herbacea 25
Salix spp. 30
Salmon 77, 138
Salmo salar 77
Salmo trutta 72
Sambucus nigra 128
Sand Eels 63, 71, 72, 169, 178, 184
Sand Gaper 107
Sand Martin 127
Sand Mason Worm 102
Saturnia pavonia 126

Index

Saxicola torquata 114
Scale Worms 68, 99
Scampi, see Dublin Bay Prawn
Scilla verna 123
Scomber scombrus 72
Scorpion Spider Crab 68
Scots Pine 132, 166, 174
Scutch Grass 113, 169
Sea Arrow Grass 118
Sea Aster 113, 114, 118, 119, 128, 178
Sea Buckthorn 25
Sea Campion 113, 114, 128, 188
Sea cucumbers 68, 99
Sea Gooseberry 57
Sea Ivory 123
Sea Lettuce 101
Sea lilies 8, 14
Sea Mayweed 115, 128, 137, 169, 188, 190
Sea Milkwort 118
Sea Pink 114, 118, 123, 178, 195
Sea Plantain 114, 118
Sea Purslane 118
Sea Slater 87
Sea Slug 63
Sea Spurrey 115
Sea squirts 92, 100
Sea Trout 72, 77, 137
Sea urchins
Sedges
Sedge Warbler 120
Sedum anglicum 112, 123
Sedum acre 123
Serrated Wrack 89, 96, 98, 101
Shag 74, 88, 190
Shanny 5, 96
Shelduck 145-147, 151, 165, 170-172
Shore Mussel 88, 101, 159, 193
Short-eared Owl 119, 163
Shoveler 146, 151, 162, 165
Silene maritima 113, 114
Silverweed 130
Six-spot Burnet Moth 124
Skate 72
Skylark 127
Slavonian Grebe 74
Slender Sea Pens 70
Small Copper 131
Small Heath 133
Small Mountain Ringlet 30
Smooth Porcelain Crab 68, 99
Smyrnium olusatrum 128
Snipe 121, 137, 146, 162

Snow Bunting 119
Snowy Owls 24
Song Thrush 127
Sorex minutus 88, 114
Sparrowhawk 149
Spartina anglica 120, 121
Speckled Wood 131
Spergularia marina 115
Spergularia media 115
Sphagnum 30, 136
Spindle Tree 130
Spiny Squat Lobsters 68
Spiny Starfish 61
Spiral Wrack 95
Spring Squill 123
Squat Lobster 99
Starling 114, 127, 149, 155
Stercorarius parasiticus 189
Sterna sandvicensis 73, 171
Sterna hirundo 74, 171
Sterna paradisaea 74, 171
Sterna dougalli 171
Stoat 125, 134, 150
Stonechat 114, 127, 135
Storm Petrel 75
Strangford Hermit Crab 68
Strawberry Worm 102
Sturnus vulgaris 114
Succissa pratensis 124
Sueda maritima 114, 118
Sula bassana 75
Sunstar 66
Swallow 177
Swans,
   Mute 137, 146, 150, 156, 171, 172
   Whooper 142, 145, 146, 150-152, 156
Sycamore 128, 133, 135
Tachybaptus ruficollis 88
Tadorna tadorna 145, 146, 171
Talpa europaea 29
Taxus 25
Teal 137, 143, 146, 152, 156, 160, 162, 165, 178
Terns, 130, 194
   Sandwich 73, 171, 177, 185-188, 193
   Common 74, 171, 177, 185-188, 190, 191
   Arctic 74, 171, 177, 185-188, 190, 191
   Roseate 171, 186-188, 190
Teucrium scorodonia 125
Thalarctos maritimus 24
Thongweed 96
Thrift, see Sea Pink
Thyme 123

Thymus praecox 123
Ticinosuchus 17
Tope 72
Tree Sparrow 134
Trifolium spp. 124
Triglochin maritima 118
Trilobites 8, 11
Tringa nebularia 146, 154
Tringa totanus 143, 146, 171
Tubulanus annulatus 68
Tubularia indivisa 63

Tufted Duck 146, 151, 170, 171
Turbellarians 103
Turciops truncatus 79
Turdus iliacus 134
Turdus philomelos 127
Turdus pilaris 134
Turnstone 139, 142, 145, 146, 153, 154, 161, 162, 178
Twite 119
Two-spotted Sucker 96
Ulex europaeus 124